Metal Scrappers and Thieves

Benjamin F. Stickle

Metal Scrappers and Thieves

Scavenging for Survival and Profit

Benjamin F. Stickle
Criminal Justice Administration
Middle Tennessee State University
Murfreesboro, Tennessee, USA

ISBN 978-3-319-57501-8   ISBN 978-3-319-57502-5   (eBook)
DOI 10.1007/978-3-319-57502-5

Library of Congress Control Number: 2017948083

Cover illustration © Richard Carlton / Alamy Stock Photo
Cover Design by Thomas Howey

Printed on acid-free paper

This Palgrave Macmillan imprint is published by Springer Nature
The registered company is Springer International Publishing AG
The registered company address is: Gewerbestrasse 11, 6330 Cham, Switzerland

To Amy

FOREWORD

Crime is a social problem that permeates society and brings harms to many and benefits to some. People suffer the consequences of crime every day, in every community. Similarly, others commit crimes daily, many times without notice or seemingly without significant consequences. Luckily, at this point in time, we know a fair amount about crime, criminals, victims, criminal contexts, and a host of other related issues. What we know about these issues, however, tends to be about particular types of criminal offenses. We have well-developed bodies of literature about homicide, robbery, burglary, theft, assault, sex offenses, drug offenses, and even many criminal offenses at misdemeanor levels. We even have a decent understanding about the basics of many "less common" or less studied crimes such as white-collar offenses, arson, and even "victimless" crimes such as prostitution. However, even with these bodies of knowledge, we are far from understanding crime and those involved in it in any comprehensive way. As Ben Stickle makes clear in *Metal Scrappers and Thieves: Scavenging for Survival and Profit*, we still have a lot to learn about crime, and for some crimes, we are literally still just scratching the surface of possible knowledge.

In this book, Stickle focuses unprecedented attention on a form of crime that is largely unrecognized, rarely discussed or addressed in policy and law, seemingly impossible for law enforcement agents to stop, and yet also a form of crime with very large, expensive, and potentially fatal consequences. The theft of metal—often from functioning, fully useful, and sometimes very important settings and locations—is a heretofore unstudied form of criminality. In bringing this form of crime to light, Stickle has ventured into a previously untapped social world and provides a solid understanding of the importance, means, actors, and challenges of engaging in metal theft. While certainly far from a "final answer" to the problem, Stickle's study of metal theft and thieves provides a strong foundation to build upon, for both future scholars and policymakers.

One of the important contributions of this book is the theoretical insight that the criminal offense of metal theft is a derivative and spin-off of an analogous yet significantly different activity and subculture: metal scrapping and metal scrappers. Through an examination of how scrappers function, who they are, how they interact with other scrappers and the broader community, Stickle delivers an understanding of the rhythms and flow of a subculture that exists in plain sight, yet is unrecognized and generally unacknowledged. What is intriguing here is not the actual actions or characteristics of scrappers, but instead how this subculture provides the foundation from which a criminal subculture arises. Scrappers are very different from metal thieves, yet they operate in ways that reflect the broader subculture. Rather than seeing scrappers and metal thieves as interchangeable and equally harmful, Stickle shows us how thieves are typically graduates of scrapping. How and where in this evolution crime emerges is the focal point of interest for those interested in arresting this form of expensive criminality.

However, it is not just advances in our understandings of crime and subcultures that Stickle provides readers with valuable insights. He also provides readers with a quintessential example of the value of qualitative research. Through interviews, interactions, and simply "being with" scrappers and thieves, Stickle shows how qualitative methods can provide rich, detailed, contextualized, and (hopefully) policy-informing knowledge. In the pages that follow, readers will experience the best that qualitative research methods have to offer: an objective, thorough, and well-explained and well-supported explanation for serious, understudied social problems.

In the end, *Metal Scrappers and Thieves: Scavenging for Survival and Profit* provides an excellent example of the application of qualitative methods, how to study a world that has not been previously examined, and how to find the ways that subcultures spawn yet other, sometimes criminal, subcultures. This is a study of more than just a simple form of theft. This is a study of a hidden world where regular people engage in regular activities that sometimes evolve into criminal activities. Crime is everywhere in our society, and as Stickle clearly illustrates in *Metal Scrappers and Thieves: Scavenging for Survival and Profit*, sometimes most of us are unaware and uninformed of some of the very serious forms of crime in our midst. After reading this book, readers will be enlightened to a hidden-in-plain-sight set of subcultures and a form of criminality that costs us all in ways we never realize.

Richard Tewksbury
University of Louisville

Acknowledgments

I acknowledge the impact and dedication of my family. I am especially indebted to my wife, Amy. Without your love, support, sacrifice, and encouragement I would not be where I am today and words cannot express my gratitude or indebtedness. To my children, Emery and Caroline, who motivate me to excel and provided much-needed relaxation and enjoyment between scrapping expeditions. To my parents, Dr. and Mrs. Fred E. Stickle, who fostered and encouraged my intellectual and moral development.

I thank Dr. Richard Tewksbury, for his guidance, encouragement, and patience; his teaching and mentorship allowed me to see academia in a new and exciting way. I also thank Dr. Patricia Gagné, Dr. Deborah Keeling, and the late Dr. Eric McCord for their assistance, insight, and guidance in this project. I am especially indebted to my undergraduate professor and mentor, Chief Donald G. Hanna, who pushed me to master my profession (and myself), as well as Dr. Susan Warner, who inspired me with her lectures to actually *see* the world. I also thank Dr. Jackie Sandifer, who gave me my first chance and encouraged me to keep going.

I also acknowledge the influence of current and past members of the criminal justice profession, who taught me the skills and character crucial to faithfully serve others and who stood by me while keeping the peace. Specifically, I acknowledge the positive impact by Doug Chisholm,

Don Parvin, Bill Randolph, Kenny Betts, Robert Hansen, Curt Clark, Josh Hughes, Ernie Steff, Robert Kitchen, and Brett Pitchford.

Finally, I acknowledge those who opened their lives to me for this book. I will not mention you by name, but your willingness to take me into your lives, your homes, your businesses, and your world has allowed me to write this book.

CONTENTS

LIST OF FIGURES

PART I

Introduction to Scrapping

CHAPTER 1

Introduction

1.1 WHY SCRAPPING AND METAL THEFT?

In 2011 I left a successful career as a sheriff's deputy to pursue a career in academia. Moving several hours away, I took a teaching position at a small regional university and began taking classes toward a PhD in Justice Administration. In an effort to save all the money I could, my wife and I purchased a foreclosed home. The only problem was that it needed work—a lot of work. I began the task of remodeling the home, and one of the first things to go was an old hot water heater that had rusted out. A few days into the remodeling I learned, based on the volume of large trash items by the roadways in the neighborhood, that bulk waste pickup was on that day. I began to roll the rusted hot water heater down the driveway toward the curb. As I did so, a small hatchback with the Pop-A-Lock company logo stopped and backed up to where I was standing. A young man in his early 20s rolled his window down and said, "Hey, do you want that?" I looked questioningly at him and slowly said "No" with a confused expression on my face. He replied with an ecstatic "Great!" and asked if I could make sure no one else took it while he did a job several houses down. "I'll be back in, like, five minutes," he said in an animated tone. Before I could even agree, he sped down the street.

© The Author(s) 2017
B.F. Stickle, *Metal Scrappers and Thieves*,
DOI 10.1007/978-3-319-57502-5_1

I stood there for a few minutes before he pulled back up and jumped out. Together we surveyed his small company hatchback and the large 80-gallon hot water heater. As he opened the passenger door and laid the seat as far back as he could, he asked if I would help him load it. I agreed, as he opened the hatchback. Together we slid the tank through the hatch and along the back of the laid-back passenger seat until the rusted bottom rested on the seat and began to leak red stained water all over it. Concerned that it was moving around too much, he then proceeded to seatbelt it into place. With great enthusiasm, he said, "Thanks so much, bro" and raised his hand for a fist bump. Still a little bewildered at the situation, I bumped fists with him, and he quickly got into his car and sped away. I stood there on the sidewalk, looking as he drove away with the hot water heater protruding from the trunk, shook my head, and thought, "What just happened? Am I missing something? How much is that worth?"

Throughout the several months of remodeling, I often thought of that experience. Particularly since I noticed that anything I set out by the roadway that was metal disapeared within a few hours. On my way to and from the hardware store, I began to notice trucks with signs advertising junk removal and, as I walked to class in the inner city, I saw countless individuals with bikes or carts loaded down with cans and other types of metal. Needless to say, the frequent observance of "scrappers" continually reminded me of that first experience. As I continued in my studies, I found that there was almost no research on scrappers or a rising crime type: metal theft.

When it came time to identify a major topic of study, I told a professor about the encounter with the hot water heater and mentioned the issues surrounding metal theft. He instantly seized on the subject and suggested I explore that with an ethnographic study. As I drove home that evening, I passed a truck making rounds in a neighborhood searching for scrap metal and thought of the times I had interacted with scrappers as a police officer and the few times I had been involved in a metal theft case. As I continued home, my mind swirled with questions: Who are scrappers? Why do they do it? How do they know where to find the metal? Are they all thieves? I realized that the incident with the hot water heater years ago had been the jumping off point in a journey—a journey that would take my skills as an investigator and communicator and transform them from investigating crimes and prosecuting criminals to understanding crimes and developing methods to prevent crime. The opportunity to convert these skills previously used behind the badge and exploit them to help understand people on the other side was tantalizing to me. And nearly before I knew it, I was trolling alleyways in my car, walking the inner city streets picking up metal, hanging out at recycling centers, and watching thieves steal metal.

What began as a confusing experience many years ago has morphed into an exciting journey with scrappers and metal thieves. This book is the culmination of my field research and provides the first known ethnographic study of scrappers and metal thieves. This work is not based on statistics or structured interviews on a college campus; rather, the study took place on the streets and in the homes of scrappers. It is an in-depth look into the lives of those who are all around us and are either unnoticed or are a societal curiosity.

1.2 Recycling in the United States

Recycling is a big industry worldwide, and particularly in the United States. The US recycling industry has a significant impact on the economics of the country. According to a report by the independent consulting firm John Dunham and Associates (2013), the industry recycled more than 135 million metric tons of material and directly employed nearly half a million persons with another 325,000 jobs indirectly supported by the scrap industry. In addition to persons employed, the recycling industry also generates more than $87 billion in annual economic activity. This economic activity accounts for 0.55% of the US gross domestic product (GDP), making it of similar size as the milk, aircraft, and cosmetic industries. The industry and its employees create approximately $10.3 billion in federal, state, and local tax revenue annually. Also, exports produced by the scrap industry are among the largest in the country, providing additional jobs and tax revenue. Scrap metal accounts for 65% of the material volume of the industry and 75% of economic activity.

However, the recycling industry is rarely involved in seeking out and acquiring items to be recycled. Rather, finding and obtaining items often falls to others who bring old, discarded, or broken items to recycling centers so that the unwanted items can enter the recovery cycle (recycling) and emerge again as new products. Because recovering material from existing products is often significantly cheaper than utilizing raw materials to create new products, companies are willing to pay for used metals, plastics, and papers. The result is that many forms of waste and unwanted material have monetary value. Some items (such as paper and plastics) have a very low value, while materials such as copper possess a very high value.

The most commonly recycled materials in the United States are various types of metals (e.g., copper, iron, steel, aluminum, brass) (John Dunham and Associates, 2013). These metals are sought primarily for economic reasons, such as availability, size, and value. Frequently, metals are readily found within the built environment, are easily broken down into manageable sizes, and maintain a high resale value. Copper, for example, in 2015, traded as high as $4.58 per pound on the Commodities Exchange Market. This means that a five-gallon bucket of scrap copper pipes or wires could bring $160 (assuming approximately 40 pounds at a purchase price of approximately $4 per pound).

Recycling centers—or scrap yards as they are often called—serve as the broker between individuals and businesses who collect recyclable material and industries who are willing to purchase and reuse the materials. Collecting materials for recycling occurs through several methods such as reclamation of large items (e.g., ships), curbside recycling, or industrial by-products and waste. Additionally, some individuals search for, collect, and recycle materials as a means of income. It is these people that this book examines. While individual recyclers may not recycle the most material by tonnage, the number of people who search for, collect, and recycle materials appears to be growing and is an area of societal curiosity.

The interest is demonstrated anecdotally by the dozens of books (both self-published and by academic publishers) providing how-to guides on metal recycling, YouTube channels dedicated to scrapping, blogs, websites, TV documentary and reality shows, online tutorials, and even smartphone apps that cater exclusively to metal recycling. Many of these resources are devoted to attracting and educating individual metal collectors on the methods, tools, and techniques of the trade. Further, many of the TV shows, documentaries, and books highlight the unusual and unfamiliar stories and behaviors of a subculture of scrappers that function within the mainstream of society.

Image 1.1   A scrapper on the way to the scrap yard

Unfortunately, the behaviors demonstrated in the media are often mis-understood by general society. For example, it is likely that the reader of this book has observed scrapping vehicles, such as an old dilapidated truck driving slowly down the streets in a neighborhood, pausing to load an old refrigerator from the curb onto an already precarious tower of old metal.

Image 1.2   Unstable metal towering above the car arriving at a scrap yard

Others may have seen handmade signs posted along highways advertising, "Free Junk Removal," "I want your scrap metal," or "We'll haul your clunker away." Just observing these individuals, their vehicles, their advertisements, or interacting with them reveals that these persons are part of a societal subculture with its own jargon, culture, and behavior.

Image 1.3   A typical sign advertising junk removal

As with many other subcultures, society is fascinated by, leery of, and often alarmed by, these individuals.

Unfortunately, there is little scholarly documentation, knowledge, or understanding of scrappers and their activities in society. Perhaps the two best-known exceptions are Jeff Ferrell's *Empire of Scrounge* and *Cash for Your Trash* by Carl A. Zimring. In *Empire of Scrounge*, Ferrell (2006) examines those who live on the edge of society by joining them in scrounging through trash. During his eight months of fieldwork living off profits from society's waste, he recounts fellow scroungers and discusses the effects of the waste prone culture of the United States. This book provides a fascinating and insightful look into individuals who live on the edge of society, and in particular, several persons in his book could be considered Subsistence Scrappers (see Chap. 3). However, Ferrell's work is not primarily geared toward metal recycling, but rather focuses on individuals who live off waste in general. Zimring's book, *Cash for Your Trash*, chronicles the historical origins of recycling and its rise in the United States from marginalized peddlers to large recycling corporations. Zimring provides a foundation to understand the historical significance of metal recycling and the industry that has developed around it. However, it does not focus on the individual scrapper either. While both of these texts provide insight into and historical context to scrapping, there is limited knowledge gained about those who scrap.

In fact, the colloquialism "scrapping," most commonly used to identify the activities of this subculture, is not adequately defined. Therefore, to establish a clear understanding, scrapping is described in this book as "the act of regularly collecting fragmented, damaged, or discarded metal items, which are no longer useful or have not maintained their original value, to recycle them for financial profit." Similarly, the term "scrapper" refers to an individual who participates in scrapping or is more generally used to identify individuals within this subculture.

The term "scrapper" is often heavily laden with latent connotations. As Zimring identifies, these connotations are often cynical, prejudiced, and rooted in historical caricatures that teeter between accurate portrayals and biased stigmas. Interestingly, cultural changes in the latter part of the twentieth century have made recycling a virtue. What was once thought of as an unsanitary and dangerous activity has been transformed, becoming fashionable and viewed as a righteous activity. However, scrappers (those who routinely recycle for profit) often continue to suffer negative stereotypes. Despite this, the number of scrappers seems to be increasing, especially during the recent US financial recession coupled with a rising price for secondary (scrap) metals and growing demand. Historical analysis by Zimring (2009) confirms that as the demand for the metal increases, so does the presence of scrappers.

1.3   Metal Theft: The Emerging Crime Type

As the price and demand for metals rise, the country seemed to experience an increase in the theft of metals. Metal theft occurs when individuals steal metal that is not theirs and sell it for the value of the metal.

As the worldwide demand for copper and other metals began to increase in the mid-2000s, the United States experienced an "epidemic" (Berinato, 2007, p. 1) of metal theft, which the FBI claims, "Threatens U.S. Critical Infrastructure" (FBI, 2008, para. 1). The Council of State Governments identified metal theft as "one of the fastest-growing crimes in the United States" (Burnett, Kussainov, & Hull, 2014, p. 1). Metal theft took a prominent position in news media with tantalizing stories of brazen metal thieves stealing everything from grave markers, sports bleachers, bridges, miles of electrical wire, air conditioners, aluminum siding from occupied homes, cars, electrical transformers from energized lines, lead roofs from churches, and manhole covers.

Moreover, many of these stories identified the significant damages and inconveniences related to the thefts. For example, while the cost to replace a lead roof stolen from a historic church is high, the damage to the floors and pews from exposure to the elements may cost many times the expense of replacing the roof. These collateral costs often account for thousands of dollars of repairs beyond the value of the metals stolen. Also, news stories often cited the inconveniences resulting from metal theft, such as hospitals running on generators due to power failures or a whole city left stranded when an entire bridge crossing a river was stolen.

Unfortunately, there is no national, state, or regional clearinghouse with data on the prevalence of metal theft or the harms associated with it (Burnett et al., 2014). Rather, the limited information known about metal theft has primarily been collected from anecdotal stories, industry reports, and a few individual cities collecting data. This limited information that is available indicates that metal theft is a costly and dangerous trend. For example, Indianapolis, Indiana, reported a 57% increase in metal theft between 2008 and 2013, which accounted for 10% of all property crime in the city and cost the residents an estimated $16.6 million (Whiteacre, Terheide, & Biggs, 2014). The Rochester Police Department in the state of New York investigated 585 burglaries related to metal theft from 2008 to 2010, which caused considerable damage to residential structures (Posick, Rocque, Whiteacre, & Mazeika, 2012). The Metropolitan Pima Alliance (2014) reported that Tucson, Arizona, suffered $2 million in damages related to metal theft in 2012, while Phoenix, Arizona, estimated losses exceeding $30 million, and had reports of over 3000 air conditioner units stolen in 2011 alone.

In addition to local government, several industries such as utilities, transportation, and insurance providers keep limited statistics on thefts related to metal theft. For example, the theft of beer kegs is estimated to cost the industry more than $45 million annually (Fischer, 2008). The National Insurance Crime Bureau (NICB) reported a 3000% increase in catalytic converter theft in 2008 (Stanfill, 2008) and an 81% growth in insurance claims related to metal theft from property between 2009 and 2011 (Kudla, 2012). The NICB also identified nearly 70,000 insurance claims related to metal theft between 2006 and 2013. The Arizona Department of Transportation announced that it suffered half a million dollars in costs related to metal theft in 2007 and 2008 (Schoenfelder, 2009), while the California Department of Transportation spent approximately $50 million in repairs since it began tracking metal theft (CTC & Associates LLC, 2013). The Electrical Safety Foundation International (ESFI) (2009) estimated that metal theft costs utility companies approximately $80 million annually.

What can be concluded from these reports is that metal theft has significantly increased in recent years and left in its wake significant costs associated with the stolen metal as well as collateral damages to structures, harming individuals and society as a whole (Bennett, 2008; Kooi, 2010; Kudla, 2012). As the rate of metal theft increased, the public began to notice, and when these statistics were combined with the media reports, state and local governments began to scramble to "do something." Despite the lack of knowledge on the topic, and specifically on metal thieves themselves, legislative bodies responded by increasing controls over recycling companies. In some cases, legislators strengthened or created laws unique to metal theft and the collateral damage often occurring in conjunction with thefts. As a result, within a few years, all 50 States had strengthened existing laws or adopted new legislation to combat metal theft (Burnett et al., 2014).

However, metal theft continued to be an issue and was seemingly unabated by new laws and efforts to curtail it (Rinehart, 2015). The failure of legal efforts is likely due to the lack of knowledge about metal theft and metal thieves. The few studies on metal theft have only examined the prevalence, trends, places, rates, and other quantitative questions. Additionally, these examinations have been limited in scope and jurisdiction. Until researchers begin to understand and learn about scrappers, and in particular about metal thieves, efforts to reduce metal theft will be futile.

REFERENCES

Bennett, L. (2008). Assets under attack: Metal theft, the built environment and the dark side of the global recycling market. *Environmental Law and Management, 20,* 176–183.

Berinato, S. (2007, February). Red gold rush: The copper theft epidemic. *Chief Security Officer.* Retrieved from http://www.csoonline.com/article/2121953/loss-prevention/red-gold-rush--the-copper-theft-epidemic.html?page=1

Burnett, J., Kussainov, N., & Hull, E. (2014). *Scrap metal thefts: Is legislation working for states?* Lexington, KY: The Council of State Governments.

CTC & Associates LLC. (2013). *Design practices and products for deterring copper wire theft.* Sacramento, CA: Caltrans Division of Research and Innovation.

Electrical Safety Foundation International. (2009). *Copper theft baseline survey of utilities.* Rosslyn, VA: Electrical Safety Foundation International.

Federal Bureau of Investigation. (2008). *Copper thefts threaten U.S. critical infrastructure.* Washington, DC: FBI Criminal Intelligence Section.

Ferrell, J. (2006). *Empire of scrounge: Inside the urban underground of dumpster diving, trash picking and street scavenging.* New York: New York University Press.

Fischer, T. (2008, February). Trouble brewing. *Scrap Magazine.* Retrieved from http://www.isri.org/news-publications/scrap-magazine/scrap-articles/trouble-brewing

John Dunham and Associates. (2013). *Economic impact study: U.S. based scrap recycling industry (2013)—Executive summary.* Washington DC: Institute of Scrap Recycling Industries, Inc.

Kooi, B. (2010). *Theft of scrap metal: Problem-oriented guides for police series. Guide No. 58.* US Department of Justice, Office of Community Oriented Policing Services.

Kudla, J. (2012). *Data analytics forecast report: Metal theft claims and questionable claims from January 1, 2009 to December 31, 2011.* Des Plaines, IL: National Insurance Crime Bureau.

Metropolitan Pima Alliance. (2014, December). *Metal watch task force.* Retrieved from http://mpaaz.org/advocacy/metal-theft-task-force

Posick, C., Rocque, M., Whiteacre, K., & Mazeika, D. (2012). Examining metal theft in context: An opportunity theory approach. *Justice Research and Policy, 14*(2), 79–102.

Rinehart, J. (2015, February 20). *Indianapolis sees surge of metal thefts.* ABC 6. Retrieved from http://www.theindychannel.com/news/local-news/indianapolis-sees-surge-of-metal-thefts

Schoenfelder, J. (2009). *Options for reducing copper theft* (Vol. 657). Arizona Department of Transportation.

Stanfill, J. (2008). *Catalytic converter thefts*. Des Plaines, IL: National Insurance Crime Bureau.

Whiteacre, K., Terheide, D., & Biggs, B. (2014). *Research brief: Metal thefts in Indianapolis October 1, 2011–September 30, 2013*. Indianapolis: University of Indianapolis Community Research Center.

Zimring, C. A. (2009). *Cash for your trash: Scrap recycling in America*. Piscataway, NJ: Rutgers University Press.

CHAPTER 2

Approach

Qualitative research often focuses on the exploration and examination of social settings and groups to recognize and understand behavior, with an emphasis on meanings, traits, people, locations, interactions, and experiences (Tewksbury, 2009). Those goals are precisely the purpose of this book, which is an attempt to recognize and understand scrappers and metal thieves. Moreover, scrappers are often viewed with disdain by general society, and metal thieves engage in criminal behavior, which makes a study of this population even more problematic. Consequently, this study's methodology involves conducting an exploratory study using an ethnographic approach.

Ethnography: "A research method that places researchers in the midst of whatever it is they study. From this vantage, researchers can examine various phenomena as perceived by participants and represent these observations as accounts" (Berg & Lune, 2012, p. 225). Ethnographic techniques have been successfully utilized in other difficult to reach, delinquent, or criminal populations (see Copes & Tewksbury, 2011; Gagné, 1992; Jacobs, 1999, 2000; Tewksbury, 1990; Wright & Decker, 2011) and are appropriate for this study as well.

2.1   DATA COLLECTION METHODS

Throughout this study, I conducted field research of scrappers and metal thieves by observing their behaviors, participating in scrapping activities alongside them, and conducting unstructured interviews. Notably, these

© The Author(s) 2017   15
B.F. Stickle, *Metal Scrappers and Thieves*,
DOI 10.1007/978-3-319-57502-5_2

three techniques were rarely mutually exclusive, but rather coalesced into a singular effort. In many situations, I observed scrappers' actions while I was participating in dumpster diving and conducting an interview at the same time.

2.1.1   Observation

One of the main tasks of any ethnographic research is to conduct field observation. Observing a group or an individual in their environment allows the researcher to gain insight into the behavior, routines, and connections within the group. This experience is vital to not only build

Image 2.1   After a long day of walking in scrapping in the inner city

understanding and knowledge, but also to prepare the researcher to enter and participate in the culture and activities of those studied. The first method I used to observe scrappers and thieves involved watching the search for metal from vehicles along streets and alleys, especially when the local waste service was collecting bulk waste. I contacted local waste service providers and obtained a list and maps of when and where these bulk pickups occurred. I spent hours traveling these routes in the early morning and late into the evening to observe scrappers and thieves in public as they searched for metal. After spending considerable time watching these actions at the recycling centers and along the streets and alleyways, I felt I had gained enough knowledge to begin incorporating participation.

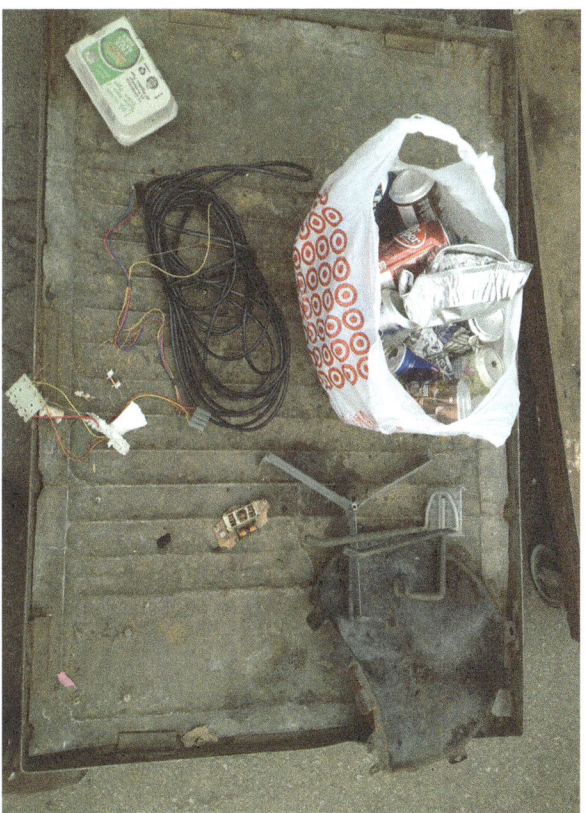

Image 2.2   I was not as successful as other scrappers, only earning $1.75

2.1.2   Participation

In addition to observation, I also participated alongside scrappers. In social science research, the role of participant as observer allows the researcher to engage in the activities of the group (Adler, 1990; Howell, 1972). The type of participant observation used in this study is commonly referred to as moderate participation, whereby I sought to maintain a balance between insider and outsider so that I could remain objective, while moderately involved (see DeWalt, DeWalt, & Wayland, 1998; Schwartz & Schwartz, 1955).

Unfortunately, my first efforts of participation proved less than fruitful. While driving through neighborhoods to observe scrappers pick through trash by the side of the road, I decided that this would be a good entry point for participation. Thus, as I saw a scrapper looking through trash or a dumpster, I would approach and begin to do the same and casually engage the scrapper in conversation. Unfortunately, as I was soon to learn (see Chaps. 3 and 4), scrappers are conflict adverse, and the scrapping cultural norms call for avoidance of other scrappers. In other words, when I joined a scrapper looking through a pile of metal by the side of the road, I would receive a confused and slightly annoyed look, followed shortly by the scrapper leaving to search the next pile of garbage. Thus, this technique of locating scrappers helped to develop an understanding of how scrappers worked in that particular setting, but did not net any significant interview contacts.

However, as the study progressed, I utilized what I was observing and the knowledge I had gained to participate in other scrapping activities. Gradually, I learned how to interact with scrappers and sought opportunities to assist scrappers in their routine tasks and to observe metal thieves. Opportunities to participate consisted of helping scrappers unload vehicles at recycling centers, traveling with scrappers and thieves in search of metal, and working with recycling center staff. This participation enhanced the information flow and provided a deeper understanding of the interactions, techniques, relationships, and meanings within the subculture.

2.1.3   Interviews

As I became more comfortable participating in and observing scrapping activity, I began to incorporate conversations, and specifically questions about scrappers and thieves. The questions were unstructured in style, with the persona of an outsider eager to understand the interviewees' experiences and lives. This method of conducting interviews allows for a natural extension of participant observation and provides a depth of understanding and valuable insights, and advances knowledge within culturally grounded contexts (Tewksbury, 2009).

Since every situation I encountered was different, I would often begin with a casual conversation about the weather or a sports game. I would then segue into an interview by saying, "Hey, I am writing a book about scrappers, people don't seem to understand them, can you tell me a little about what you do?" or "Everyone thinks scrappers are druggies and thieves, is that true?" While this was often the jumping off point for a meandering conversation about scrapping, I maintained a mental checklist, so to speak, of topics I wanted to cover before the end of the interview. In some circumstances, these conversational interviews lasted hours; at other times, some refused to talk with me, and some interviews were short and did not cover all the topics I desired.

As mentioned previously, many conversations occurred while inside a dumpster, on the side of the street, when unloading metal from vehicles, while helping to separate metals, over a cup of coffee, or while sharing a cigarette. In this way, the open-ended questions of an unstructured interview allowed me leeway to direct the interview in much the same way as a conversation between acquaintances. Based on reflexive listening, I provided follow-up questions, constructed new questions, pursued clarification, and guided the interviewee toward the targeted knowledge base. The greatest insight occurred when the interviews transpired in a location where the respondent was comfortable. Thus, many discussions took place in the recycling center—or yard, as it was often called—along the respondents' favorite scrapping route, inside their home, and at other convenient locations. The familiar environment provided an optimal setting for dynamic situations where the interviewee provided specific information I had not requested, thus allowing rich details to arise within the conversation.

For example, while I was interviewing Toney, a metal thief, he offered to walk me to the building where he was burglarizing and stealing metal. We walked for nearly a mile through a blighted urban area and talked. During our conversation, I used the houses and buildings we passed as examples to engage him in conversation about metal theft and to discover what he found attractive or detracting about individual buildings. These observations brought about significant details about the methods used to commit metal theft that may not have naturally developed while conducting standard interviews in a controlled environment.

2.2   FIELD SETTING AND RESPONDENTS

2.2.1   Field Settings

Nearly all the interviews of this study were conducted in the field, although a few individuals preferred to correspond by phone or email. In addition to interviews, I spent approximately 150 hours of fieldwork observing and participating in the scrapper subculture. The interviews, observation, and participation occurred between October 2014 and April 2015 in Indiana, Kentucky, Oregon, and Tennessee. Also, a few phone or email interviews took place with scrappers and thieves who operated in Florida, New York, and Ohio.

The majority of the fieldwork took place at recycling centers. Recycling centers proved to be a stable location, which provided a suitable reason to have extended and repeated contacts (see Chap. 4: At the Yard). This setting is particularly useful for locating thieves because one of the few outlets to exchange stolen metal for money occurs at recycling centers. Thus, nearly all metal thieves, at some point, journey to a recycling center and mingle with scrappers and staff while selling their metals.

2.2.2   Demographics

I conducted unstructured interviews with 55 individuals who were either scrappers or metal thieves. The interviewees were predominately male (85%) and had a racial makeup of 69% white, 29% black, and 2% Hispanic. The ages varied from as young as 15 to an individual in the late 80s, although the most common age range were people in their 30s (24%) and their 50s (22%). While 27% had some college experience, 25% were unemployed, and the rest were employed, retired, or receiving government assistance.

In addition to interviews with scrappers and metal thieves, I interviewed 12 current and retired law enforcement officers who were active, or had been active, in investigating metal theft. These interviews were also unstructured, and provided an example of how metal thieves are commonly viewed by law enforcement agents as well as identifying what knowledge officers had of metal thieves and how they operated.

As the research progressed, broad taxonomies of scrappers and metal thieves were identified (see Chap. 3). Thus, several months into the study I began to seek particular taxonomies as the foci. This technique is often called a purposive sample. A purposive sample occurs when researchers use their unique knowledge and expertise about some group to select subjects who represent this population (Small, 2009). Specifically, as I began to immerse myself into the subculture of scrappers and metal thieves, I gained knowledge and insight into the behavior and characteristics of scrapping; this expertise allowed me to identify and engage those suspected of metal theft. For example, while observing activity at one scrap yard, I noticed the metal that Toney was selling seemed suspicious. He was selling mixed metals such as an old burglar alarm, an electrical breaker panel door and cover, a security light, a few short pieces of copper pipes, and various types and lengths of electrical wire. Due to the extensive observation and knowledge I had recently developed, I suspected he was involved in metal theft and worked to establish a relationship with him. This experience enhanced the identification, contact and understanding of metal thieves, as they were easier to recognize and engage, with relevant questions targeted at their activities.

2.3   GETTING IN

Of course, the knowledge necessary to identify metal thieves would be worthless if I could not be where they were, specifically scrap yards. Therefore, I had to *get in* to the locations that scrappers and metal thieves frequented and gain assistance from others who were already there. *Getting in* often refers to the techniques used to obtain access to a setting and the participants (Friedman, Bell, & Berger, 2003). *Getting in* with difficult to reach and often stigmatized groups, such as scrappers and especially metal thieves, can be very problematic for an outsider, as the routines, rituals, and argot (specialized language) are unknown. However, gaining entrée into delinquent or criminal populations is certainly possible (see Cromwell, Olson, & Avary, 1991; Ferrell, 2006; Jacobs, 1999; Wright & Decker, 1994; Wright, Decker, Redfern, & Smith, 1992).

To enhance the likelihood of successful fieldwork, I followed the observation phases suggested by Howell (1972). These initial stages included gaining knowledge by visiting the research sites, establishing rapport, and gaining support from gatekeepers. Since there were very limited academic publications on scrappers to use as a resource, it was vital that I develop entry-level knowledge through other means. Thankfully, there is a plethora of other resources: self-published autobiographies, how-to manuals, informative websites, online group discussion forums, YouTube videos, smartphone apps, and even reality TV shows and a documentary movie on scrappers. I studied these resources before any attempts at *getting in*, so that I could express a degree of limited knowledge with the interviewees and be better prepared to understand and participate in the unique aspects of this subculture.

2.3.1   Establishing Trust

Successfully getting in within scrappers and with metal, thieves required me to alter my appearance, manner of speech, and—to some degree—actions. While I never purposely presented myself as a scrapper or thief, I quickly realized I would have to change certain things about my appearance and habits before I was accepted and trusted. Growing up in a middle-class neighborhood, attending graduate school, and having served as a police officer in the past, I was not well positioned to interact with those on the edges of society.

The differences and difficulty were evident from day one, as many scrappers either asked me if I were an undercover cop or were skeptical of me. After carefully considering what I could do to alter the quick judgments I was receiving, one thing was clear: I could not pretend to be a thief or a scrapper. Presenting myself as one of them would be a deception that they would clearly see through, and I would burn any bridge I had built. Rather, I had to be approachable by reducing suspicion, yet remain an outsider.

To gain trust and get into the culture, I had to change my physical characteristics and speech patterns. I grew my hair out until it was shaggy and unkempt, and allowed my goatee—which seemed to scream "undercover cop"—to grow into a full beard. I wore old clothes, often frayed around the edges, and always carried cigarettes. Smoking was a technique I learned as a cop and, seeing its success during tense situations, I employed it often in the field. There is something nearly magical about offering a smoke to a stranger; it propels a relationship to the next level and speaks volumes without saying a word. Many scrappers and a few metal thieves visibly relaxed while we shared a smoke.

I also began to mirror those I was interviewing and, when appropriate, would swear, use slang, and in general match their cadence and demeanor. I also carried a small amount of money to purchase lunch or drinks as I talked with people. As my hair grew and my clothing became increasingly worn from scrapping, my ability to talk like many of the scrappers and thieves also increased. As I became not so different from them and was more relaxed in the settings, those interviewed likewise became more and more comfortable around me, signaling that I was being accepted and was getting into the subculture.

2.3.2   *"I Checked You Out": In with the Thieves*

Developing rapport and trust within any group under study can be difficult; this is particularly the case for many of those I interacted with and only saw once. I usually had one chance to make a good impression and present myself as someone to be trusted, knowable enough to discuss scrapping but still an outsider. In other words, I had to walk and talk as if I knew what I was doing, while not appearing to know too much. Either extreme could compromise the information I was getting.

One particularly challenging group to get in with was metal thieves. Since their activities were illegal, they clearly had a reason to distrust me. However, on several occasions, it was clear that I was able to get in and had garnered a reputation for myself as someone who could be trusted and was "not a damn cop." One incident in particular clearly illustrates how I was accepted among thieves.

I had received a referral that Zach was involved in metal theft. I attempted to talk with him several times, but he always blew me off. Then, one night I contacted him by phone, and he agreed to meet me at his house. Near the end of the interview when I asked for an introduction to other metal thieves, the interview took an interesting turn.

> *Zach:* I know several [thieves] that I can talk to and see if they will talk [to you], but I have to tell them you ain't no damn law. They won't talk to nobody.
> *Interviewer:* I understand
> *Zach:* Well… see, at first, when I first met you, something didn't fit, you know, I was kinda curious about you. To tell you the truth about it, I didn't know you from Adam. I didn't know if you were the law or undercover or whatever. But I finally found out when I started checking.
> *Interviewer:* Oh yeah?
> *Zach:* Yeah, I checked you out!
> *Interviewer:* Did you? What did you find out?
> *Zach:* I ain't gonna lie to you. I got hold of some friends of mine in [a city where I had completed a significant amount of fieldwork,

about two hours away from the location of this interview], and they know you.

Interviewer: Do they? (rather confused and surprised)

Zach: They know you 'cause, you do stories of people up there.

Interviewer: So I've talked with some of your friends up there?

Zach: Yeah, a couple of homeless people.

Interviewer: Are you talking about Jared?

Zach: Yeah, he used to live here with me!

As we continued to talk, I told Zach of how I had helped Jared carry metal to the scrap yard without asking for anything in return. He then replied:

> See that's what got me interested in you, I come home and told him [over the phone] about [the first time I met] you, he [Jared] said, 'He's a cop', and I said well, if he is, he ain't gonna get me on nothing. But you kept coming by, and eventually, I called Jared [again]. [This time] he said he knew you better and you weren't a cop, and I knew I could trust ya.

To my surprise, my efforts to establish rapport in a city over 100 miles away had translated into trust with another thief. I had always surmised that Jared was involved in metal theft, but he always claimed that he was "clean." Given the strong connection with another thief, I had my doubts. Regardless, at least among these individuals, I was *in*.

2.3.3   *"Oh, There's the Gun": In with the Yards*

While I was making headway among some thieves and scrappers, I also had to get in with the scrap yards. I had conducted some observation at several yards and sold small amounts of metal to a few more. However, hanging around the yards and talking with people required building a relationship with the yard staff. Therefore, I sought the cooperation and support of gatekeepers. Gatekeepers are individuals in positions to grant or deny access to a research setting (Feldman et al., 2003). The gatekeepers within a recycling center are, typically, the owners. These individuals provide opportunities for an insider's perspective through their relationships, background, and access to scrappers. I approached many recycling center owners, presented the purpose of the study, and sought permission to contact and interview customers who arrived at the yard. In most cases,

I received permission. Similar to the example with Zach, many yard owners provided verbal references to other yard owners asking them to support my efforts.

In most cases, the owners introduced me to the staff who could serve
as guides. Guides are persons found within the subculture who can assist
the researcher in gaining credibility and making contacts (O'Leary, 2005).
The most common guides were yard managers or scale operators who handle the day-to-day functions of customer service. Interestingly, the majority of scrap yard employees either scrap on the side or were very active

Image 2.3   The author spending time at a local scrap yard. This was early in the
study, before I had grown out my hair

scrappers before yard employment. These individuals therefore know and understand metals, their value, persons who sell them, and other important factors related to scrapping. Scale operators and yard managers served two important roles: first, by lending their credibility to my activities to the other staff and customers; and second, by helping to connect me with scrappers.

As I spent time at the yards, I engaged the workers in conversation. I explained what I was doing and listened to their stories. I was quick to lend a hand when they were busy, often helping to weight metal or help customers find the pay window, back up their vehicles, or unload trailers. These activities furthered my rapport with the staff. It became apparent that I was getting in, as the staff would tell me, "You'll want to talk to her," or tell a customer, "If you have some time, talk with this guy, he writing a book, he's cool." These introductions spawned many interesting conversations, as the guides knew the backstories and activities of many customers who visited the scrap yard.

The clearest example of being *in* with the yard staff occurred several months into the study when I was asked to watch the pay window.

Employee: Can you do me a favor?
Interviewer: Yeah.
Employee: Watch that door; guard it with your life! I've got to go out to my car for a minute.
Interviewer: My life? [Said sarcastically, as I nodded my head in agreement to the request]
Employee: Yep! Oh! [Pausing halfway out the door] There's the gun. [Pointing to a handgun just out of sight of the public view and next to the money drawer]

In a short time, I had proven myself as helpful, knowable, and trustworthy; and in this case, I was considered part of the staff. Undoubtedly, I was *in*.

2.4   DATA ANALYSIS AND PRESENTATION

During each observation, participation, and interview, I kept field notes and audio recordings of the interviews I had conducted and the thoughts and perceptions I had during my fieldwork. These audio recordings were transcribed verbatim, a pseudonym assigned to each participant, location

names and identifying references altered, and other identifying information changed. After these steps, the audio recordings were destroyed.
These measures ensured anonymity of the participants.

Once the fieldwork was complete, a content analysis was performed.
A content analysis is a careful, detailed, systematic examination and interpretation of social communication to identify patterns, themes in events
and persons and meanings, which produce a cohesive representation of
the subject (Neuendorf, 2002; Tewksbury, 2009). I examined both manifest and latent content. Manifest content consists of data that is physically
present and quantifiable (e.g., written transcripts), whereas latent content
includes identifying symbolism underlying the manifest data (e.g., the verbal and physical mannerisms during the interview recorded in field notes)
(Berg & Lune, 2012).

While conducting the content analysis, I used the social anthropologi-
cal approach identified by Miles and Huberman (1994). This method
helps detect and explain the way people operate in particular settings and
how they understand, make sense of, respond to, and function in their
day-to-day life. Researchers who have spent considerable time within a
given subculture and have participated (directly or indirectly) within the
subculture are best suited to conduct this type of analysis.

I began by completing a thorough reading of all field notes and interviews
to identify themes developed during the study, and to determine new themes
previously unrealized. After the initial reading, I continued to examine the
data in two stages. The first stage was open coding, which is the process of
thoroughly reading, identifying, and extracting themes and topics discovered within the data (Strauss, 1990). The second reading or focused coding
resulted in additional coding and linking of data. Collections of codes containing similar content are grouped together to form concepts. Recurrent
concepts are linked together and coded again to narrow the focus and develop
categories that provide rich details on the subjects and situations. Focused
coding allowed for a structured, repeated, and detailed examination of the
data that identified concepts and categories (both latent and manifest) which
confirmed, disproved, and proposed theories of the subculture studied.

In the chapters that follow, I present the scrappers and metal thieves'
views of what they do, how they view themselves and others, and why they
behave as they do. Each chapter focuses more narrowly taking the reader
from a general overview of scrappers to a detailed analysis of metal thieves.
Chapter 3 defines scrappers and metal thieves and provides a taxonomy of

each type of scrapper and metal thief. Chapter 4 examines the subculture of scrappers and explains how metal thieves do not maintain the same codes and norms of the scrapper subculture. As we move into the remainder of the book, the focus narrows exclusively to metal thieves. Chapter 5 provides a short primer on metal and an analysis of the historical events leading to the rise of metal theft, and also identifies the prevalence and costs associated with metal theft. Chapter 6 examines the state of the literature on metal theft, identifying what we know and what we think we know. Chapter 7 examines metal thieves' motivations. Chapter 8 discusses the techniques and methods used by metal thieves, including how places attract theft, the process of completing the theft, and techniques to sell the stolen goods. Chapter 9 looks at the ways thieves collaborate and how they learned the techniques and methods of metal theft, while Chap. 10 identifies the direct, indirect, and internal controls that affect thieves. Finally, the implications of the findings for future research, criminal justice policy, and criminological theory are discussed in Chap. 11.

Chapters 7 through 10 draw heavily on quotes drawn from scrappers and metal thieves. These quotes are chosen to illustrate the concepts discussed and were selected because they generally represented the perceptions presented by the scrappers or thieves on a particular topic. Great effort has been taken to avoid bias or embellish preconceived notions. While this method of content analysis may have some moderate weakness, it is well suited for this study; an exploratory examination into a hidden subculture and emerging criminal activity that remains largely unknown in the current literature.

REFERENCES

Adler, P. (1990). Ethnographic research on hidden populations: Penetrating the drug world. In E. Y. Lambert (Ed.), *The collection and interpretation of data from hidden populations* (pp. 96–112). Rockville, MD: National Institute on Drug Abuse.

Berg, B. L., & Lune, H. (2012). *Qualitative research methods for the social science.* New York: Pearson.

Copes, H., & Tewksbury, R. (2011). Criminal experience and perceptions of risk: What auto thieves fear when stealing cars. *Journal of Crime and Justice, 34*(1), 62–79.

Cromwell, P. F., Olson, J. N., Avary, D., & W. (1991). *Breaking and entering: An ethnographic analysis of burglary.* Newbury Park, CA: Sage Publications.

DeWalt, K. M., DeWalt, B. R., & Wayland, C. B. (1998). Participant observation. In H. R. Bernard (Ed.), *Handbook of methods in cultural anthropology* (pp. 259–299). Walnut Cree, CA: Altamira Press.

Feldman, M. S., Bell, J., & Berger, M. T. (2003). *Gaining access: A practical and theoretical guide for qualitative researchers.* Walnut Creek, CA: Altamira Press.

Ferrell, J. (2006). *Empire of scrounge: Inside the urban underground of dumpster diving, trash picking and street scavenging.* New York: New York University Press.

Gagné, P. L. (1992). Appalachian women violence and social control. *Journal of Contemporary Ethnography, 20*(4), 387–415.

Howell, J. T. (1972). *Hard living on clay street: Portraits of blue collar families.* Prospect Heights, IL: Waveland Press, Inc.

Jacobs, B. A. (1999). *Dealing crack: The social world of streetcorner selling.* Boston: Northeastern University Press.

Jacobs, B. A. (2000). *Robbing drug dealers: Violence beyond the law.* Piscataway, NJ: Transaction Publishers.

Miles, M. B., & Huberman, A. M. (1994). *Qualitative data analysis: An expanded sourcebook.* Thousand Oaks, CA: Sage Publications.

Neuendorf, K. A. (2002). *The content analysis guidebook* (Vol. 300). Thousand Oaks, CA: Sage Publications.

O'Leary, Z. (2005). *Researching real-world problems: A guide to methods of inquiry.* Thousand Oaks, CA: Sage.

Schwartz, M. S., & Schwartz, C. G. (1955). Problems in participant observation. *American Journal of Sociology, 60,* 343–353.

Small, M. L. (2009). How many cases do I need? On science and the logic of case selection in field-based research. *Ethnography, 10*(1), 5–38.

Strauss, A. L. (1990). *Qualitative analysis for social scientists* (2nd ed.). New York: Cambridge University Press.

Tewksbury, R. (1990). Patrons of porn: Research notes on the clientele of adult bookstores. *Deviant Behavior, 11*(3), 259–271.

Tewksbury, R. (2009). Qualitative versus quantitative methods: Understanding why qualitative methods are superior for criminology and criminal justice. *Journal of Theoretical and Philosophical Criminology, 1*(1), 38–58.

Wright, R. T., & Decker, S. H. (1994). *Burglars on the job.* Boston: Northeastern University Press.

Wright, R. T., & Decker, S. H. (2011). *Armed robbers in action: Stickups and street culture.* Boston: Northeastern University Press.

Wright, R. T., Decker, S. H., Redfern, A. K., & Smith, D. L. (1992). A snowball's chance in hell: Doing fieldwork with active residential burglars. *Journal of Research in Crime and Delinquency, 29*(2), 148–161.

Taxonomy of Scrappers and Metal Thieves

The first place to start in any research is to define who is being studied and identify what they are doing. Unfortunately, this task is rather difficult for scrappers and metal thieves. As is discussed in the previous chapters, not a lot of research exists on scrappers; in fact, the term scrapping may have never been defined. Also, there has never been a study of metal thieves, and the occurrence of metal theft on a massive scale is understudied. Thus, to explore these topics, definitions need to be established. Once those definitions are set, information about scrappers and metal thieves needs to be examined to determine the differences, if any, between scrappers and metal thieves. This chapter takes up these tasks by establishing working definitions of scrapping and metal theft. Next, it develops a taxonomy of scrappers and metal thieves, providing a method to classify and better understand the individuals active in this subculture.

3.1   What Is Scrapping and Metal Theft?

Scrappers and metal thieves are at the center of significant media, public, and government attention. Unfortunately, the absence of sound knowledge has created a void, which is often filled with unfounded claims and misguided assumptions. Part of the reason for this has been a lack of

© The Author(s) 2017   31
B.F. Stickle, *Metal Scrappers and Thieves*,
DOI 10.1007/978-3-319-57502-5_3

defined concepts. Despite the significant usage of the word and knowledge of the existence of persons who recycle, scrappers and what they do (scrapping) does not appear to be defined in literature or popular culture. Therefore, it is necessary to define this concept to provide the foundation for this book, and for future research in this area.

Scrapping: "the act of regularly collecting fragmented, damaged, or discarded metal items, which are no longer useful or have not maintained their original value, to recycle them for financial profit."

Even more concerning is the confusion of how to define metal theft. While a detailed discussion of the term metal theft occurs in Chap. 7, the term is defined here to avoid confusion. Throughout the book, the definition developed by Kevin Whiteacre, Terheide, and Biggs (2014) is used to establish what is and is not metal theft.

Metal Theft: "the theft of item(s) for the value of the constituent metals."

3.2   Taxonomy

Now that the actions of scrapping and metal theft have been defined, it is necessary to identify those who are involved in these activities and determine how they can be grouped for easier study. This process is often referred to as developing a taxonomy. Establishing a taxonomy is an important first step in studying any group of people. Classifying those involved in scrapping and metal theft will allow for an advanced conceptualization of characteristics, motivations, shared experiences, and relationships with others within and without the taxonomy.

However, developing a taxonomy of scrappers and metal thieves is challenging, primarily because there is no typical scrapper. The divisions cannot be made by age, employment, gender, education, race, or any commonly attributed obvious distinction. However, careful observation of actions and interactions within the setting of recycling centers and while searching for metal, combined with unstructured interviews, make it possible to distinguish and categorize scrappers and metal thieves. Based on hundreds of hours of participation, observation, and conversation, I identified five varieties of scrappers and metal thieves: Subsistence Scrappers, Scrapping Professionals, Professionals who Scrap, Philanthropic Scrappers, and metal thieves. It is important to note that this taxonomy does not include individuals who occasionally recycle metal. While this group is large, the infrequent nature of their recycling habits, the low level of knowledge and skill, and the lack of their involvement or identification within the subculture indicate that they should not be included.

3.2.1   Subsistence Scrapper

Subsistence Scrappers are individuals who scrap to earn money for necessities or to supplement a limited income that does not currently meet their needs. These individuals are unique in three important aspects: their financial and social status, their scrapping technique, and their motivation. Subsistence Scrappers consist of a wide variety of individuals, including a broad range of ages, both genders, and many races. In fact, one of the only consistent characteristics is limited financial means.

A sizable portion of Subsistence Scrappers is homeless. Some are transient and may live in and out of shelters or homeless camps, while others reside in substandard housing conditions. Subsistence scrappers appear typically disheveled with a rather unkempt, poor appearance, often attributed to a lack of income and resources for personal hygiene. This unkempt appearance should not be confused with Scrapping Professionals who are often dirty from engaging in their scrapping work.

Subsistence Scrappers locate their metal by searching along roadways, alleys, through public recycling bins, garbage cans, dumpsters, and other places near the boundaries of society. Subsistence Scrappers do not have vehicles, as they are too expensive; rather, they rely on walking or bicycle riding to search for metals. Subsistence Scrappers often use grocery carts, backpacks, or other devices adapted to carry larger volumes of metal to the recycling centers. Likewise, Subsistence Scrappers who utilize bicycles often have bags of metal slung across handlebars or have adapted milk crates or other containers to hold their metals.

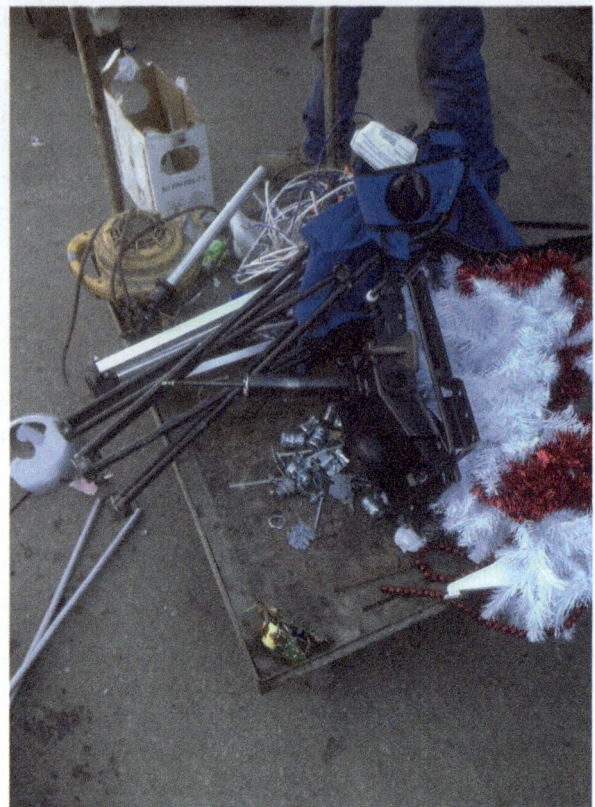

Image 3.1 A typical collection of metal from a Subsistence Scrapper

Subsistence Scrappers are located exclusively within urban environments. This location is a necessity as these scrappers rely on walking or bicycles to fulfill all their daily needs, including scrapping, purchasing food, and other daily activities. Subsistence Scrappers have a detailed knowledge of the area in which they live and scrap, and often have contacts that provide metal for them (e.g., neighbors who save cans for them, businesses that allow them to procure metal from their dumpsters).

Money is a particularly important motivating factor for Subsistence Scrappers, especially as nearly all Subsistence Scrappers interviewed were unemployed or disabled. A small minority held intermittent part-time jobs, but the majority were unemployed. Consequently, most considered scrapping their "work." For example, when discussing scrapping, Subsistence

Scrappers frequently used language consistent with those employed full-time in general society: "Not bad for a day's work" or "I work every day." Roughly half of the Subsistence Scrappers I interviewed received limited government assistance; for example, many received food stamps and a few received housing aid or disability income. Very few Subsistence Scrappers received more than one government assistance benefit, leaving them to make up the financial shortfall by scrapping. While nearly all the Subsistence Scrappers discussed their enjoyment of scrapping, the lack of financial funds from traditional jobs or government assistance substantiates that most Subsistence Scrappers are scrapping out of a genuine financial hardship, in other words simply to "survive."

Scrapper Profile

Name: Nathan
Taxonomy: Subsistence Scrapper
Location: Southeast—Inner City
Age: 59
Race: Black
Education: High school or less
Employment: Government Assistance
Experience: Nathan rode what he called "the scrap bike," which was an old dilapidated bicycle that had been converted to help carry metals found along his rides. He lived in government housing and received government financial assistance, and said that he scrapped to "make ends meet, you know the food stamps aren't enough, besides I don't get them until next week and I am struggling." Nathan was incredibly animated when speaking and struggled to focus on the questions I asked him. When I told him I was writing a book on scrappers, he informed me he was also working on a book, *Hard Times in America*. I asked him if he was having a hard time in America, and he reared back, waving his arms, and said, "Man! It's all in the book!"

While Subsistence Scrappers experience significant and consistent needs of financial resources to survive (e.g., pay bills, purchase food), the most frequently mentioned motivation for scrapping was cigarettes and alcohol.

Image 3.2   Nathan's scrap bike

Coming in next were mentions of bills, followed by food. Lastly, a small portion of Subsistence Scrappers said that using or purchasing drugs was a motivator. While buying cigarettes and alcohol was often discussed, only a limited number of Subsistence Scrappers chatted about being alcoholics or appeared to be intoxicated on a regular basis; rather, alcohol—and especially cigarettes—were more often used as entertainment, for enjoyment, and for social interaction.

Similarly, drug usage was mentioned on occasion, but their use was often understood to occur after other needs had been met. For instance, Chad, who lived in a dilapidated house and received $400 a month in retirement, explained scrapping this way: "Yeah, I get cigarette money, cigar money, spice (synthetic marijuana) money, blunt money." It is important to distinguish between the primary motivation for scrapping in conjunction with alcohol, drugs, and tobacco usage and the motivation to support passive consumption. In other words, these activities (e.g., tobacco, alcohol, minor drug usage) do not drive scrappers to scrap; rather, they are often the reward for scrapping. For example, very few of the Subsistence Scrappers appeared intoxicated while scrapping and none said they scrapped exclusively to support a drug, alcohol, or tobacco habit; rather, those vices were used afterward for enjoyment, celebration, and socialization.

Subsistence Scrappers were the largest group studied, comprising 33% (n = 18) of those interviewed, but accounted for the smallest volume of metal recycling. This small number is primarily due to limited transportation that prohibits them from acquiring large amounts of metal or searching in a wide area. Their lack of mobility also means they must compete for the same scarce resources. All the Subsistence Scrappers interviewed expressed a strong disdain for metal thieves, often proudly telling the interviewer, "I'm not no thief," and explaining that being known as a thief would limit the contributions others in the neighborhood voluntarily give them, as well as cause issues at the recycling centers if employees knew they were selling stolen metal.

Overall, Subsistence Scrappers are struggling to survive financially, are generally homeless, and walk or bicycle to public areas looking for small amounts of metal (e.g., usually cans), so they can earn a small amount of cash each day. Their income from scrapping rarely rises above $10 a day and is often used for food, cigarettes, or alcohol. Subsistence Scrappers are rarely involved in theft, and have an extensive knowledge and relationship with each other and those in the community.

3.2.2   Scrapping Professionals: Metal Is Their Business

Scrapping Professionals are individuals who acquire the majority of their income from scrapping. Scrapping Professionals accounted for 20% (n = 11) of the present study, but are perhaps the most publicly visible and immediately recognizable of all scrappers. They tend to have vehicles loaded down with the various metals or have signs advertising their services. Once again, there are few physically distinguishing characteristics of this group. During this study, I observed and interviewed individuals of all age ranges, different education, and differing backgrounds, races, and genders. The only immediately apparent consistency is the tendency to work in teams. Scrapping Professionals are unique in two important ways: first, scrapping occurs on a massive scale, utilizing vehicles and a wide array of tools; second, Scrapping Professionals tend to make the majority of their income from scrapping.

Scrapper Profile

Name: Nicole
Taxonomy: Scrapping Professional
Location: Southeast—Urban City
Age: Mid 50s
Race: White
Education: Some College
Employment: Unemployed
Experience: I met Nicole at a scrap yard after an introduction by the staff. She was a frail woman with apparent back pain, who drove a nice SUV, had a small dog with her, and appeared to be middle class. All of the metals she was selling had been neatly separated by type (e.g., copper, aluminum) and she knew the yard staff well. She explained that when she and her husband lost their jobs during the economic downturn, she turned to scrapping in 2009, adding that "I made $10,000 off free stuff on the side of the road (the first year) and it blew my mind. So, I started looking at it in a totally different aspect." She described how she used a trailer to search along the streets and eventually made connections with companies such as mechanic shops, and would dumpster dive for old car parts. When I asked her why she continues to do it years later, she said, "To survive, the environment, extra money, it keeps me busy." She also described the addiction to the rush of finding metal. Recently her techniques had changed from large items using a trailer to smaller items; she described how, in addition to driving around the city, she would stop at yard sales looking for metal items. For example, she said, one time "I spent $5 at a garage sale on a stack of silver plate trays, turned out to be [actual] silver, 79 ounces, so I got [over] $700 off that."

   The techniques to locate metal take many forms that are unique to this taxonomy. Frequently, Scrapping Professionals proactively advertise services for junk removal. In these cases, individuals call and may even contract with a Scrapping Professional to clean out junk from a field, garage, basement, or other location. Scrapping Professionals respond and evaluate the quantity and quality of metal. If satisfactory, the Scrapping Professionals may charge a fee to remove the unwanted material, dispose of non-metal items, and then scrap the remaining metal items. Another enterprising technique involves advertising junk car removal. Each car transferred to a recycling center can bring between $300 and $700 depending on vehicle weight and the price of the metal.

   Scrapping Professionals also drive routes searching for metal items set out for public waste collection. In fact, one city garbage collection manager described how most cities no longer publish bulk trash pickup dates due to scrappers who flock to the area. He explained how he and many others view scrappers as an annoyance and as individuals who reduce the recycling income for the city. Despite the efforts by governments to keep bulk waste collection routes and dates secret, Scrapping Professionals

Image 3.3   A scrapping professional advertising his services

typically know the days and locations of the bulk waste collection in the areas they work, and have a detailed understanding of when it is the best time to scrap. A few of the Scrapping Professionals would also collect non-metal items for resale at consignment stores, flea markets, or by other means. However, this was uncommon.

Nearly all the Scrapping Professionals interviewed discussed having partnerships with local businesses which supply them with discarded metal. These companies would call or set aside metal for Scrapping Professionals to pick up on a regular schedule. While not often discussed in any detail, it appeared this technique was a source of a substantial portion of income for Scrapping Professionals. They also have a detailed knowledge of the streets and byways of their community. Likewise, these individuals may operate in urban and rural areas since they possess transportation (e.g., pickup trucks), which allows them to search quickly and efficiently for larger metal items.

Scrapping Professionals are also unique in the degree to which they process metals. To gain the maximum value for their metal, many have a detailed process of separating each type of metal. For example, several Scrapping Professionals interviewed bring metal items back to their house to break them down by metal type (e.g., copper, brass, aluminum). Once the metals are completely separated and their storage space is near full capacity, they will take particular types of metal to the recycling center. However, not all Scrapping Professionals operated this way. Some merely piled scrap metal (e.g., bed frames, couch springs, washing machines) onto their truck and sold the mixed metals for a bulk price. Selling bulk mixed metal results in a reduced price per pound of material but is less time-consuming.

The Scrapping Professionals interviewed maintained a keen social awareness of each other; many of them communicated at the recycling center and off the yard, typically through cell phones. These relationships were often friendly, and Scrapping Professionals were quick to assist other Scrapping Professionals when the need arose.

Unlike the other taxonomies, Scrapping Professionals relied nearly entirely on their abilities as scrappers to earn an income. Therefore, they had to be skilled at scrapping to succeed, which required knowledge of how to scrap, adequate transportation and tools, along with social contacts with other Scrapping Professionals. Many Scrapping Professionals' skills, abilities, knowledge, and means of transportation represented the highest achievement in the scrapping subculture and were frequently envied by Subsistence Scrappers.

3.2.3   Professionals Who Scrap: In the Course of Business

Professionals who Scrap are individuals who collect and recycle metal during the course of their regular employment, and made up 15% ($n = 9$) of the present study. Examples include plumbers, electricians, window or siding installers, HVAC (heating, ventilation, and air conditioning) workers, construction laborers, maintenance employees, and auto mechanics. This category does not include individuals who just place a dumpster on their property to collect scrap metal for wholesale to a recycling center on a monthly contract. Rather, Professionals who Scrap are individuals who put forth a concerted and purposeful effort to locate, identify, and remove valuable metal during the course of their work. Examples may include a plumber who carefully collects the remaining copper piping parts until the end of the workweek or a construction laborer who sells excess rebar (steel reinforcement used in concrete) after a completed job.

The demographics (e.g., age, race, gender) of Professionals who Scrap is similar to the demographics of persons who work in related fields (e.g., electricians, plumbers). Thus, this category tends to include younger white males, but is not exclusive to these individuals. Moreover, the diversity of employment circumstances and metal materials observed in this group are extensive. Many professionals who scrapped visited the recycling center on a regular basis. Others would come as soon as they had an adequate load of metals to recycle, which varied extensively based on the type of metals collected. For example, electricians often accumulated small amounts of wire over time and frequently visited once every few weeks or so, whereas a siding installer might come after each siding job.

The type of metal sold varied according to the field the individuals worked in and might include copper, aluminum, steel, brass, iron, lead, and other uncommon metals. While Professionals who Scrap usually only bring in the metals they have available at work, occasionally they also bring metals they discover in the course of their job. For example, a HVAC installer may observe a refrigerator on the side of the road while on the way to a job and include the refrigerator with the other metal he collected from old HVAC equipment. On occasion, Professionals who Scrap work as a team and split the money earned, but more often than not function as individual scrappers. The funds made by these people often augment their current income and are frequently seen as "fun money," spent on food, alcohol, or other luxuries.

Image 3.4   These items are unused sheet metal a Professional who Scraps brought to the yard for sale after a job

Other than thieves, this group shared the least similarities with the others categories of scrappers. Some of the Professionals who Scrap claimed to have been a Subsistence Scrapper or a Scrapping Professional before being hired in their current employment status. However, once employed full-time, none scrapped extensively outside of work. Conversely, many scrappers in other taxonomies learned the value of metals by being a Professional who Scraps and transitioning into another category when employment in the field ended.

Professionals who Scrap are viewed as highly valued, trustworthy, and frequent customers at recycling centers. Their status as practitioners in

Image 3.5   Used window frames a Professional who Scraps brought during a large remodel job

an industry removes a significant portion of the suspicion and stigma that is associated with other types of scrappers and thieves. This trust may or may not be warranted, as Professionals who Scrap tend to move in and out of the taxonomy dependent on employment and have significant opportunities for theft and dishonest activities when working with metal for a customer. Professionals who Scrap are difficult to study, as they tend to stop at yards only between jobs, have little time for interviews, and are not as social as others within the scrapping subculture.

Scrapper Profile

Name: Corey
Taxonomy: Professionals who Scrap
Location: Southeast—Urban City
Age: 22
Race: White
Education: Some College
Employment: Employed Full-Time
Experience: Corey worked in the concrete industry as a laborer;
               however, he often recycled the remaining metal
               items after a job. When I met him, he had just used
               his truck and another co-worker's trailer to recycle
               ten tons of extra rebar from a job, earning more
               than $1000 (before splitting with his co-worker).
               He explained how it was common in the industry
               to overbid on a job and then scrap what was
               remaining, saying, "In the end, the company
               doesn't want it. When something is already bought
               and paid for, they won't send it back and don't
               know what else to do with it." In his particular
               case, because Corey had a truck, the foreman
               selected him to get the metal and used the com-
               pany skid loader to load it. When I asked Corey
               about the ethics of the situation, he said, "It's not
               a very well-tracked business for sure, I mean a
               $1000 in steel and you think somebody might
               want to ask a question."

3.2.4   Philanthropic Scrappers

Philanthropic Scrappers are individuals who scrap primarily for philanthropic purposes. Included in this category are people who raise money
for charities, utilize the funds earned from scrapping for humanitarian
efforts, or collect metal to give to others—often to Subsistence Scrappers,
to help them make a living. Philanthropic Scrappers tended to be older

and male; however, this was not always the case among those interviewed. These scrappers were by far the smallest category observed during the present study, accounting for only 8% ($n = 4$), yet unique enough to garner its distinction.

The way the funds earned from scrapping are used is the primary difference between Philanthropic Scrappers and other types of scrappers. Philanthropic Scrappers exclusively use the funds on behalf of others, rarely, if ever, keeping any of the profits for themselves. These individuals may scrap in the course of their work, actively search for metal on a routine basis, or collect metals from friends and family. Regardless of their collection methods, the goal is singular: to help others.

For example, one individual, who, before retirement was a Professional who Scrapped, organized people within his church to help him search for discarded metal as a fundraiser for a new building addition. He and a team of church members sought, collected, processed (separated the different types of metals), and recycled metal totaling over $35,000 during a four-year period. Another Philanthropic Scrapper collected metal in the course of his work as a maintenance employee and used the funds to purchase materials utilized for a religious course he taught at the local addiction and homeless shelter. Other Philanthropic Scrappers collected cans to contribute to groups conducting can drive as fundraisers (e.g., Boy Scouts). A few assisted Subsistence Scrappers by collecting metal for them, providing a vehicle to help haul a hefty find, and on occasion selling the metal on behalf of a Subsistence Scrapper if a government issued ID was necessary for the sale.

Scrapper Profile

Name:	Justin
Taxonomy:	Philanthropic Scrapper
Location:	Southwest—Inner City
Age:	Mid-60s
Race:	Black
Education:	Unknown
Employment:	Full-Time

Experience: I met Justin at a scrap yard and engaged him in conversation. He was a jovial man and was quick to tell me that he scraps to support his ministry at a local homeless shelter and at the local jail. He explained, "I gather up what I can now, put it into the fund, when next year comes around I spend everything on books, supplies, and Bibles I use to teach classes at the jail and charity house." Justin explained that he works as a maintenance person at a local company and they allow him to keep the scrap metal. He told me, "I keep all of this stuff that is thrown away. What I do is regenerate this to teach Bible classes." As we parted, he invited me to visit his church and seemed genuinely interested in helping others through his scrapping efforts.

Regardless of the method of collection or designation of the profit, the Philanthropic Scrappers were actively involved in the general subculture of scrappers. Each of the Philanthropic Scrappers knew others who scrapped across the other taxonomies. An exception to this was metal thieves; no Philanthropic Scrapper admitted to knowing a metal thief. Nearly every Philanthropic Scrapper engaged other individuals in scrapping, either encouraging or suggesting the activity among their acquaintances or teaching others how to scrap.

3.2.5   Metal Thieves

Metal thieves are people who take metal which they have no legal right to possess and recycle it for personal gain. Specifically, for inclusion in this taxonomy the individual either admitted to or hinted at a history of metal theft. Several scrappers in other taxonomies discussed a single incident of theft or shared an experience they did not consider theft, but that would likely have met the legal definition of theft. However, the Metal Theft taxonomy is inclusive of individuals who purposely, frequently, or exclusively engage in theft to obtain metals. Of those involved in the present study, 24% ($n = 13$) were metal thieves.

This taxonomy again defies the standard measurement categories, as metal thieves vary in race, sex, age, and education. However, certain similarities help to describe common metal thieves. Nearly all metal thieves operate in teams, at least for the majority of their crimes. Teams of thieves usually develop from longstanding friendships, family, or due to romantic relationships. Occasionally, metal thieves will operate individually outside of the group; however, the majority of thefts occur in pairs or larger groups. While the relationships within the group are relatively stable, the relationships between other metal thieves are often negative. For example, metal thieves may steal metal from other metal thieves or engage in physical altercations over territory.

Nearly all metal thieves interviewed were employed full-time. In fact, many of them were employed in well-paying jobs requiring specialized skills or licenses, such as electricians, HVAC workers, general maintenance workers, or contracting workers. This employment provides several significant advantages for metal thieves. First, their background and expertise provides technical skills, knowledge, and often the tools necessary to carry out large, and technically challenging amounts of metal theft (e.g., removing miles of copper wire from energized electrical stations). Second, their regular employment often exposes them to areas with high quantities of valuable metals that are frequently unguarded. Third, metal thieves' employment status in an industry that often deals with metal can be used as a cover for their thieving activities or serves as a convenient excuse to be in possession of large amounts of metal.

Beyond the skills and knowledge developed through their employment, many metal thieves began scrapping legally, before drifting into criminality. The reasons for the drift into criminal behavior vary among thieves, but commonly involve the ease of theft, significant financial incentive, and low risk, often coupled with an initial perceived financial need. However, most metal thieves, unlike all the other taxonomies, do not enjoy collecting metal. Whereas most other types of scrappers express an enjoyment and even an addiction and excitement to scrapping, most metal thieves do not receive similar gratification for their actions; in fact, many express guilt and remorse for their actions or justify their behavior. Moreover, metal thieves do not function within the subculture of scrappers after they begin stealing; rather, they know and work with each other and no longer hold to the general norms, beliefs, mores, and codes of scrappers.

The motivation for metal theft, as with many types of crime, is money. Interestingly, very few of the metal thieves interviewed were in need of money to meet necessities; rather, many utilized the money to augment their regular employment and spent the additional funds for pleasurable activities, to assist relatives who were in financial need, or, in a minority of cases, to purchase drugs. Each metal thief who discussed drug usage as a factor identified maintaining a high or not being sick (e.g., coming down from a drug-induced high) as a motivating factor for metal theft. However, none of the metal thieves who were drug users exclusively used the profits for drug pursuits.

Some metal thieves were caught stealing, while others were arrested on unrelated charges and sent to prison. The majority of metal thieves stated that they could not stop taking metal until after a long bout in prison or jail regardless of whether the imprisonment was related to metal theft or other charges. Overall, metal thieves are unique from scrappers and share very few similarities. Their thought processes, motives, skills, abilities, and techniques are widely different.

3.3 SUMMARY

The term scrapping has been a common colloquialism for years, yet the precise definition has received little, if any, attention. The lack of definition has enhanced the difficulty of studying, or even identifying, scrappers or those who are "regularly collecting fragmented, damaged, or discarded metal items, which are no longer useful or have not maintained their original value, to recycle them for financial profit." With the establishment of this precise definition, several significant advancements are contained in this chapter.

The first advancement is an essential distinction between scrappers and individuals who occasionally recycle or do so casually at work or home. This difference is an organic division that the majority of people interviewed already knew and understood. For example, when the general population scraps an old car, takes a broken water heater to the scrap yard, or saves cans and recycles them occasionally, they do not consider themselves a scrapper. Rather, a scrapper is someone who regularly searches for and recycles old, used, or unwanted metal.

The second advancement is the identification and explanation of the diversity among scrappers. While scrappers are dissimilar from general society, they are not a cohesive group without internal peculiarities. Scrappers reflect a broad range of demographics, social characteristics, methods, and motivations that make classification necessary yet difficult. This chapter recognized four distinct taxonomies (or classifications) of scrappers: Substance Scrappers, Scrapping Professionals, Professionals who Scrap, and Philanthropic Scrappers.

Subsistence Scrappers are individuals who scrap to earn money for necessities or to supplement a limited income that does not currently meet their needs. Scrapping Professionals are people who acquire the majority of their revenue from scrapping. Professionals who Scrap are individuals who collect and recycle metal during the course of their regular employment (e.g., electricians, plumbers). Lastly, Philanthropic Scrappers are people who scrap primarily for charitable purposes. These taxonomies are unique and diverse, yet it is common for scrappers to move from one category to another throughout their scrapping experiences.

Finally, and perhaps the most significant advancement in this chapter, is the distinction between metal thieves and scrappers. metal thieves are people who take metal which they have no legal right to possess and recycle it for personal gain, which is an important division that many in society—and in law enforcement specifically—fail to observe. The danger is that scrappers may be subject to intense organizational mistrust, societal suspicion, and enhanced legal scrutiny just because they are viewed as metal thieves. Another risk of failing to discriminate metal thieves from scrappers means that scarce resources used to reduce metal theft are not targeted appropriately.

References

Whiteacre, K., Terheide, D., & Biggs, B. (2014). *Research brief: Metal thefts in Indianapolis October 1, 2011–September 30, 2013.* Indianapolis: University of Indianapolis Community Research Center.

CHAPTER 4

Scrappers as a Subculture

Regardless of whether a person is a Subsistence Scrapper, Scrapping Professional, Professional who Scraps, or Philanthropic Scrapper, scrappers share certain beliefs and behaviors that are different from the dominant culture. To varying degrees, scrappers function in and remain a part of the dominant culture, yet their actions and beliefs create a variance between themselves and general society, thus creating a subculture. Subcultures can be defined as "a separate reality through a distinct set of norms, values, mores and attitudes that contrast with those of a larger and more dominant culture" (Miller et al., 2011, p. 115). These differences often manifest themselves in a subculture's actions, norms, and codes.

4.1   The Scrapper's Code

Scrappers are always searching for metal. In some ways, the search never seems to end and is often a daily task necessary for financial support or because the activity is addictive and enjoyable. Regardless of the reason, the constant search for metal serves to enhance the divide between scrappers and general society. For example, Mrs. Jackson, whose only source of income is scrapping, stated, "I do this all the time ... I got to always look for cans. That is how I make money." Conversely, Cody and Amber said they search "all the time." When asked if scrapping was how they

© The Author(s) 2017   51
B.F. Stickle, *Metal Scrappers and Thieves*,
DOI 10.1007/978-3-319-57502-5_4

made their living, both exclaimed in unison, "No!" laughing while they explained that they both held jobs outside of scrapping. Regardless of the reason, nearly all of the scrappers interviewed indicated that they scrapped on a regular basis, with just over half suggesting something similar to Dustin's remark: "I scrap every day because I need the money."

In addition to the need to be always searching for metal, the number of persons involved in scrapping has increased dramatically. Leo recollected his experience of scrapping in 2009, saying,

> Everyone knows about it now. You know this whole scrap thing started with the way the economy was going, you know, and once the recession hit that's when everybody started doing the scrapping thing, and now it's this big, big thing.

What is more, during participant observation, I visited a recycling center on the Saturday before Christmas and found the scrapping crowd so large that it took over an hour in line to reach one of two scales, each of which was operating with a team of employees to assist in the unloading, separating, and weighing process.

After weighing the metal, I received a check because the onsite ATM that typically dispensed payment had been depleted of cash after only a few hours of operation. While this experience was atypical for me, many of the recycling center employees frequently reminisced of times in recent memory when similar long lines were the norm. Cory, a scrap yard employee, mentioned how busy the yard was when he began working for the scrap yard in 2008, saying, "... we had people lined up outside the fence and everything."

The amount of metal available to be scavenged by scrappers is, to some degree, finite. In other words, while metal products are frequently discarded and replaced (e.g., appliances, aluminum cans, construction materials), the volume of these discarded materials may not increase at the same rate or in the same place as those seeking to scrap them. This geographic and supply disparity was particularly the case during the US economic recession during the late 2000s into the early 2010s, which led to a reduction in consumer purchases and decreases in construction projects. This decrease in available scrap metal, coupled with an increase in individuals turning to scrapping as a way to augment a declining or non-existent income, significantly increased competition for scarce sources of

Image 4.1   The long line the day before Christmas stretched into the parking lot and took over two hours of waiting

valuable metal. Carlos, a scrapping professional, pointed out, "… now there are a lot of people doing [scrapping],'cause so many people are out of work. A lot of people are doing it now," and then explained that he does not scrap as often as he used to back in the 1980s and 1990s because the current competition is so intense.

Scrapper Profile

Name: Carlos
Taxonomy: Scrapping Professional
Location: Southeast–Urban City
Age: Early 60s
Race: Black
Education: Unknown
Employment: Unclear
Experience: I met Carlos at a scrap yard. He appeared worn and tired, even beyond his age. He explained how he was an active Professional Scrapper in the 1980s and 1990s and only recently returned to scrapping because of money issues. Due to injuries, he was not able to scrap as he used to, so he relied on others to collect metal items for him. He also taught his son to scrap and sends him out on a regular basis. He explained, "I use this money to buy my firewood for my house. It's some extra money that you can get for whatever, a sideline hustle."

The potential for conflicts between scrappers while searching for and acquiring scarce resources or during the theft of metals seems probable. This is especially so since scrappers as a subculture are marginalized by society, function outside the traditional physical boundaries of society (e.g., alleyways, trash bins), and frequently operate within questionable limitations of legality (e.g., trespassing, collecting metal from marked recycling bins). Therefore, it is doubtful that traditional means to establish norms or settle disputes, such as law enforcement, would be available or applicable to scrappers. In fact, the present study revealed that scrappers had developed their own subcultural norms, values, and mores. These behaviors and beliefs are described as a *scrapper's code*, which specifies acceptable behavior within the scrapping subculture. This concept was never directly labeled as such, nor are aspects of it directly explained by any individual scrappers. Rather, throughout the present study, consistent themes and concepts appeared during interviews, observations, and participation, which allowed me to identify a simple scrapper's code.

4.1.1   Territorial Issues

Locating the metal is perhaps the most apparent sphere of potential strain between scrappers and an area where norms and codes of behavior are vital for harmony within the scrapping subculture. The general methods and techniques for locating metal tend to be unique within the taxonomies and rarely cross. For example, Subsistence Scrappers tend to dumpster dive or search along alleyways for small and easily carried metals, while Scrapping Professionals tend to use vehicles to search for larger items or receive requests from customers to remove large volumes of metal. Professionals who Scrap in the course of their business locate metal while conducting their work, and metal thieves usually find metal where few others tend to tread—abandoned buildings or on private property. These separate techniques, which remain steadfastly within the taxonomies, reduce the likelihood of territorial disputes crossing taxonomy boundaries.

However, within the taxonomies the competition to find and acquire metal is high. For example, while searching for metal with Spoons, a Subsistence Scrapper, I asked him how much competition there was. He perked up, lowering his head slightly as his eyes widened and he said in a very grave drawn-out tone, "A bunch!" He went on to estimate that 30 scrapping teams are active in the small area of town where he typically scrapped. Despite the significant number of scrappers searching for the same resource, the present study revealed very few reports of arguments or conflicts related to scrapping territory. It was not that scrappers did not have routes or territories that they quasi-claimed first right to; rather, scrappers tended to operate on the axiom of *first come, first serve*, or as James put it, "I mean [the] early bird gets the worm." Similarly, Jared, who has been scrapping for over two decades, explained the unspoken code:

Interviewer: Is there a particular route that you go?
Jared: Yeah, I've got a route that I do every day. There are other people that hit it too.
Interviewer: Do you get along with them?
Jared: Oh yeah! Street people are all brothers. Now if I come down this street and I see a guy over here picking up cans, I will take a different route or the other side of the street. We have respect for each other.
Interviewer: So you kinda avoid each other?
Jared: Well, not really, if I see he has that, I will let him have that. I show him a little respect, and I hope he does the same thing for me.
Interviewer: And for the most part that happens?
Jared: Yeah!

Jared's beliefs were shared by most scrappers I spoke with, including David. When David was asked how he would respond "if someone was in [your territory]," he echoed Jared and many others, simply saying scrappers "respect each other." This typically means letting the first one in the area collect the metal unmolested, even if that is a part of a scrapper's typical route or a favorite place.

However, not all territorial encounters were solved uneventfully. Gary shared an argument he recently had with another scrapper:

> *Gary:* I get along with everybody, but one man. Because he thinks that he is the only one that can go in this certain dumpster. I mean, when you go in the dumpster, whoever comes first that's who … You know if I go somewhere and someone is already doing it, I ain't going to mess with them. [You] supposed to move on when you see somebody else. But this one man, he comes to me yesterday hollering, "You can't get in this dumpster. Man, you better get on your way, I know the owner of this dumpster, and he already told me I can go in there."
>
> *Interviewer:* So you cannot claim it unless you are there.
>
> *Gary:* Yeah, it's who was there first. But if I am there I'm not going to quit 'cause he walked up.

This situation was resolved by Gary continuing to search for metal while the other scrapper yelled and cursed, eventually departing to seek metal elsewhere. Based on the present study, it would seem territorial disputes are rare. However, in cases where a conflict does arise, the perceived norm violator, the one claiming territory, is simply told what Jennings said to another scrapper who had attempted to claim territory, "I told them 'first come, first serve … you don't own this [place], it's not yours, you don't have your name on it, and you can't tell me what to do.'"

While infrequent territorial disputes did occur, it was far more common for scrappers to share territory or even to recommend locations for other scrappers to search for metal. However, sharing territory or location secrets carried with it certain norms as well. For example, Shane said that he and his scrapping partner "shared locations, but only [with] the ones that helped us out as well." Likewise, James, a Professional Scrapper, indicated he shares location secrets with other scrappers, but expects them not to visit those places without him:

Like, for instance, say I know several scrappers, and sometimes they roll with me, or I roll with them, and they know certain places to go, and I know certain places to go, and you know when you find them at your place one day and you all ain't [scrapping] together. Okay, some get mad [and think] you should only come here with me.

Likewise, Nicole, also a Professional Scrapper, said:

> *Nicole*: I found a few people will share information, but it is very competitive. A lot of people won't share information. So, if, for example, you say, "Yeah, I know here and here" … The next thing you know, they will be there trying to pick it up.
>
> *Interviewer*: So you can share too much, right?
>
> *Nicole*: Exactly, like it's competitive!

Although there is substantial competition among scrappers to find places with consistent quantities of metal, very few scrappers experienced conflict with each other during this process. The lack of conflict is primarily due to several codes of conduct commonly held within the scrapping subculture, such as *first come, first serve* and *showing respect*. These codes, which entail conflict avoidance, ordinarily prevent or dissipate a potentially negative situation of a territorial dispute. When norms are violated by fellow scrappers, the aggrieved scrapper is quick to remind the other of the codes governing territorial claims. Finally, scrapping is a competitive process, and while the fertile territory is often kept secret, limited information is exchanged under certain circumstances, typically in a *quid pro quo* manner.

4.1.2   Not a Thief

With rising competition among scrappers, the increasing value of scrap metal, and a decrease in the supply of readily available sources of metal, it seems probable that instances of theft would increase among scrappers. After all, as the subculture already often lives and operates in the marginalized parts of society (e.g., back roads, slums, abandoned properties), it is doubtful that items missing from these areas would be noticed. However, each time I discussed metal theft within the scrapper subculture, nearly all denied ever stealing; in fact, many expressed an extreme dislike for metal thieves.

After establishing a relationship with a scrapper, in some cases spanning hours or days, I would directly ask if they were ever tempted to steal, or I might point out metal that could be quickly taken and ask if they were ever tempted. In each case, the questions or suggestions were couched in non-accusatory ways, purposely asked or staged to evaluate if the scrapper steals metal. The most common reply to this issue, topic, or idea was an emphatic, "I'm not a thief!" A few, like Chad, went on to tell me exactly what he thought of thieves, saying, "I see the motherfuckers stealing! I don't like thieves." It is important to understand that many scrappers agreed that the possibility of theft was often present, but almost none of them admitted to stealing. Consider Jared's comments, which generally represent the scrappers interviewed:

> *Jared*: I ran into a guy one time who said, "I know where we can get some copper. But we got to wait until it's dark." I said, "I don't want nothing to do with it."
> *Interviewer*: Now, why is that? What keeps you from going and stealing it?
> *Jared*: Because I'm not no crook! And I don't want it to come back on me. Because you know I like to do my shit legal.

Other scrappers cited ethics, character, honesty, or faith, indicating that they would not steal for one or several of these reasons. For example, Chad claimed that his Native American heritage prevented him from stealing:

> I am Native American, and I think it's very un-Native American to go around doing illegal shit ... I mean look how hard I go out and work for the shit I need, so why do I want to take something out of someone else's hands? I can't do that shit. I have to live with myself. I can live with that old stinky beer smell on my hands better than I can that fucking guilt of stealing off people.

Similarly, when I asked Jennings what kept him from being tempted to steal, he replied, "God, He knows everything I do, I was raised in a Christian home, so He knows everything I do ... People work hard and earned dollars for the shit they've got ... I [earn] honest money."

Scrapper Profile

Name: Jennings
Taxonomy: Subsistence Scrapper
Location: West Coast
Age: 39
Race: White
Education: Some College
Employment: Government Assistance
Experience: While walking the streets of a tiny town on the West Coast, I spotted Jennings pushing an old 60-inch tube TV down the main road in town while pulling a bike with a flat tire. He was wearing rain pants and a pullover and was filthy. He was missing all but two of his top teeth and most of the bottom ones, and talked in a loud and excited manner while constantly smoking cigarettes. Jennings traveled between Alaska and nearly every state on the East Coast, doing whatever he needed to survive. This often involved scrapping, as was the intent with the TV he had found on the curb and had pushed for nearly a mile. Jennings had been scrapping since he was 18 and claimed, "I probably made close to $90,000 in all the shit I found over the years." While Jennings certainly stuck me as the type to exaggerate a bit, he also did not strictly scrap and would often sell items to pawn stores in addition to scrap yards. He explained that he lives the way he does by choice and scraps because "it's almost like an adrenaline high because you never know what the freak you're going to find when you open that lid [of the dumpster]." As the rain picked up, he called for a cab to take him and the TV to the scrap yard. When the taxi van arrived, I worked to help him get the TV in the van, but it would not fit. So it was simply left where he had dragged it to, blocking the sidewalk. As the rain increased, he said goodbye and rode off on his bike with one flat tire to a friend's house.

A few others mentioned that their past work in the security profession and their honest nature prevented them from considering theft or even from trespassing. For example, when I asked what kept Craig from stealing, he said, "Hell! I was 20 years [in the] security office, anti-theft. I wouldn't even dream like that. I would not even walk in a man's yard unless he said I could." Likewise, Cowboy remarked:

> That is something I will not do, I will not steal nothing from nobody, I'll ask for it ... I've been that way all of my life, I was a security officer over at [a well-known professional sports complex], and I've just got it bred in my system, to be honest with people.

While the majority of scrappers cited personal convictions, honesty, or faith as reasons they do not steal, a small number mentioned the concern of being caught. As Chris commented:

> *Chris:* You know, I almost brought [to the scrap yard] a fucking [street] sign; you know one of those signs that says no left turn. It fell as I was coming up [the road], it fell right in the middle of the street.
> *Interviewer:* So why didn't you bring it?
> *Chris:* 'Cause, I would probably get in trouble. You know it. Probably get in trouble for that. I can't sell something that belongs to the state. You know.

Frank made a similar comment:

> *Frank:* ...You see people stripping houses. I wouldn't do that 'cause I am not going to jail.
> *Interviewer:* Ok, so you don't do it because you don't want to go to jail. Any other reason?
> *Frank:* 'Cause it's not legal. If it was legal, none of these houses would have copper in them.
> *Interviewer:* Is it ever tempting?
> *Frank:* Honestly, yes, it is! 'Cause you know, you are walking past an abandoned house and you know there is like $400 in that basement just sitting there. But I can't do it, 'cause I am not criminal minded like that. I'm not going to risk $400 for two years when I can make $400 in a week at [scrapping].

It appeared clear through interviews with scrappers that none of them openly supported theft. Likewise, many of them outright condemned the practice, and most of them righteously claimed, "I'm not a thief!" While the reasons for not stealing varied, the majority centered on some ethical reasoning, prior work experience, or concern of being caught. However, one thing seemed clear: scrappers, as a whole, claim they do not steal and do not like people who do. These subcultural mores ran deep throughout all taxonomies, except the thieves. Moreover, when I asked for referrals to those who are engaged in stealing, scrappers nearly unanimously commented they did not know of any. A few said they knew of several thieves, but not well enough to know details such as a phone number or name. It appears that scrappers, for the most part, have little to do with metal thieves and have adopted a code of *no theft* amongst themselves.

4.1.3   Ask First

Another question asked of nearly all the scrappers in the present study was, "How do you know what is free for the taking and what belongs to someone?" This question was designed to elicit the evaluation process of the scrapper, and his or her ability to distinguish between items that they should and should not take. To my surprise, there was great consistency in the responses, again demonstrating the norms and values that characterize scrappers. The most commonly cited method concerned situations where the items in question were in a dumpster, trash can, or set out on the curb. For example, Frank said, "But I don't take nothing unless I ask the person or it's by the trash can, you know what I am saying? If it's by the trash can, it's fair game." Another common method to determine if the item is free for the taking is "if it's broke" or if it appeared damaged and unused, which also signaled to many scrappers that the item was free to grab. The third most common description of articles scrappers considered free for the taking were items left in an alley. Antonio explained it this way:

> Well, that's, I guess, that is a personal call for an individual. If it's in the alley and there is a bunch of stuff lying around. Most people don't leave their stuff in an alley, you know what I mean? If it's in the front yard laying around, obviously it's somebody else's. But, if it's in the back alley and laying there either someone else took it and dumped it, or somebody dumped it. If it's in the alley, it counts as fair game.

When pressed further about distinguishing items "up for grabs," scrappers continuously discussed the standard code of *ask first*. For example, Frank shared an incident to demonstrate the code, saying, "I came across some copper pieces. Man, some nice copper, #1 copper! I seen it and asked the guy (property owner), 'Hey, can I have this since you are going to throw it away?' He was like 'Yeah.'" Several scrappers suggested they *asked first*, so as not to harm a relationship with the property owners who allow them to scrap on their land. Sean pointed out, "But sometimes, you know, you can go around to different places and ask people, and they give you stuff." Chad discussed his fear of being cut off from a source:

> I don't steal no cans, man; I pick them up, you know what I mean? You can steal fucking cans, but it would be stupid. If I took a bunch of cans without asking, as soon as everyone finds out they won't give me none no more.

Scrapper Profile

Name:	Sean
Taxonomy:	Scrapping Professional
Location:	Southeast
Age:	Late 40s
Race:	White
Education:	Unknown
Employment:	Self-Employed
Experience:	While sitting outside of a scrap yard I observed an old white van driving down the road, making loud sounds. As Sean stepped out and commented that it was broken but would cost too much to fix, I engaged him and his son (who is in his late teens and scraps with his dad). They described how they have contracts with some local stores to split the money on metal that the store sets aside for them to come and get. He spoke about how he makes "a little bit here and there," saying, "It's kinda tight right now." Both Sean and his son were very polite and explained how their entire family scrapped, but they were quick to leave since it was still early and there was time to get more. Unfortunately, the car would not start unless pushed, and as there was a slight rise in the street, I offered to help push, along with Sean's son, to get the van going.

When scrappers asked for permission but did not receive a response, they were split on whether no response qualified as permission. For example, if the resident was not at home or did not respond to the door, the scrapper had to make a choice to take the item in question or leave it. Craig implied that if he were not given express permission, he would not take the things because he would then consider himself a thief:

> *Interviewer.* OK, so do you ask? Like if you see something, do you go to the door and ask?
> *Craig.* I have before. But if they don't answer I'm not gonna. That is bad business. A thief is a thief. You know, I mean that is why they steal; they are a thief. And, they are always looking for something like that.

Samantha, who also trained her cousin to scrap, was less clear about the necessity of *asking first*:

> I told him you got to be careful. I have told him you can't be going on nobody's property or anything like that. He knows, when he goes scrapping or when he goes through the alleys. I told him, you got to you know if you got enough energy go up to the house and ask the guys, can you have the refrigerators in the alley, sitting there. You know it's ready to get rid of, but go ask. See if they care to get rid of it. Obviously, it's just sitting there and no one's [been able] to get rid of it. (Implying she would take it).

Regardless, a substantial majority of scrappers discussed *asking first* which was suggested not only to avoid situations where taking an item could be claimed as theft, but also because "it's what you're supposed to do" or, as Jerome said, "I don't [scrap] in someone's yard or nothing. No, I ain't no thief." Even when the item is clearly set out for trash, Nicole said, "Sometimes I even go up to the house, ring the doorbell and ask 'cause it's something that I can't believe that they are throwing away."

In addition to the consistent distaste for metal theft, many scrappers had a well-formed understanding of what was available for taking and what was not. Clear boundaries of what is available where, and when, appeared to be established across most taxonomies and in all locations of the present study. Moreover, the code of *ask first* was purportedly commonly adhered to within the scrapping subculture. Whenever doubt existed or even if something seemed too good to be true, *ask first* as a code of behavior was commonly cited as a method to keep from stealing or being accused of doing so, either of which harmed the scrappers' image within the scrapper community and among those who might provide metal to individuals.

4.2   A Cohesive Community

The scrapping subculture also maintains a fellowship within itself and displays a surprising degree of loyalty and helpfulness toward one another. In other words, most scrappers within the community are acquainted with one another and maintain those friendships while scrapping. Specifically, in situations where assistance is needed to collect metal or a scrapper is in need of personal assistance, scrappers tend to step forward and assist each other. This community cohesion and cooperative atmosphere among scrappers further substantiate that scrappers operate within a subculture, which not only has shared experiences, but also maintains values and behavioral expectations. The community among scrappers is clearly seen in three circumstances: at the yard, on the streets, and when in need.

4.2.1   At the Yard

Since scrappers work individually or in small teams dispersed across a community searching for metal, the primary time and location to develop relationships occurs at the recycling center or "yard." It is during the time spent together at the yards that scrappers tend to develop relationships with each other, as well as with the recycling center staff. While each recycling center is unique in its layout, procedures, and processes, the yards visited during the present study functioned in much the same way, and scrappers' experiences are broadly divided into four stages. Each of these stages provides an opportunity for scrappers to engage in conversation.

During the first stage, scrappers arrive on foot, bicycle, or in a vehicle and unload their metal onto the carts or into containers provided by the recycling center. This process may be time-consuming, especially when the scrapper has a large volume of metal. During this stage, I occasionally observed scrappers assisting each other to unload large items, sharing tools, or providing suggestions and insight on separating certain types of metals. When the businesses are not hectic, employees often assist customers to unload while engaging them in conversation and occasionally explaining how to separate metals more efficiently. During this stage, I met Ray, who brought in buckets of rusted and mud-covered metal, much of which was unidentifiable. As the yard staff helped him unload the metal, Ray explained how his son, who does remodeling work, called him and suggested he come by the house he was working on to get some metal from a flooded basement. Ray asked his son, "Is it worth my gas?" and his son said, "Yes." Ray looked at me with a doubtful expression, and as other scrappers gathered around, he pointed at the unidentifiable brown-colored metal and said in desperation, "And look at what I got! I told him, 'Man, you are kidding me!'"

Most of the scrappers who were present agreed that the metal may not have been worth his time and began to have separate conversations with each other, reminiscing about times they had worked to get metal that failed to pay off financially.

Image 4.2   Mud-covered metal Ray had to recycle

Image 4.3   Waiting in line at a scrap yard, many scrappers engage in conversation with each other and share techniques

The second stage begins as soon as the metal is unloaded and the scrapper wheels the carts or containers to the scales. While waiting for their turn at the scale, many scrappers carry on general conversations or briefly comment on what they or other scrappers had brought in, especially if it was a "big score," a "good lick," or a rare metal. While waiting in line for the scale during covert participation, I was engaged in conversation with Robert. As the conversation continued, I pointed out the others talking and asked about relationships at the yard. Robert indicated that most people speak with each other, telling me, "If there's a long line, I'll sit here and bullshit like I'm doing with you."

Once at the scale, the metal is separated and weighed according to type (e.g., brass, aluminum, copper). During this time, the demeanor of most scrappers changes, the lighthearted social conversations that occurred moments prior tend to pause, while scrappers intently watch their metal being evaluated, separated, and weighed by yard staff. I witnessed many scrappers subconsciously demonstrate signs of tension or anxiety during the latter part of this stage. Scrappers might tightly wring their gloves, nervously tug at clothing, purse their lips together, or impatiently sway on the balls of their feet. Most scrappers' eyes rapidly dart back and forth between the scale weight display and the metals being sorted by employees. Many of these nervous movements occurred with Ray too. During the weighing process, he intently observed the yard staff and the scale weight, eagerly waiting to learn if his efforts would pay off. While some scrappers remain fairly silent during the weighing process, others anxiously engage in superficial or nervous conversation with the staff and occasionally with other scrappers.

After weighing the metal, stage three involves collecting payment. Conversations during stage three typically occur while scrappers are waiting in line to receive payment from a cashier or an ATM. Conversations during this time are usually more lighthearted and include topics beyond metal. Depending on how happy the scrapper is with the results, their attitude could vary from relief and enthusiasm to frustration and pessimism. Regardless of the outcome, however, scrappers tended to discuss their triumph or disappointment with other scrappers casually. For instance, Ray's mud-covered metal produced a surprising amount:

Staff: $12.26, Ray. (Ray's eyes widen as he began to laugh, looking at each person standing nearby, waving his claim ticket.)
Ray (yelling): Thank you!
Interviewer: Are you happy?
Ray: Yeah! That's my gas money AND a pack of cigarettes! I'm going to enjoy this! (As Ray walks away, he shakes his head in disbelief while giggling and talking to himself) $12, I ain't never! Ha!

After payment is received, scrappers move on to an optional last stage. It is at this time that some scrappers choose to linger around the yard for a short period to socialize with fellow scrappers. For example, after receiving payment, Robert sought me out in the yard, even running, and flagging me down as I was driving away. Robert and I continued to engage in conversation, this time about things other than metal. After a short period, phone numbers were exchanged, and later Robert even offered to share his recently acquired marijuana, proudly pulling it from his pocket and presenting to me for a close inspection of its quality. While some scrappers frequently spent time talking among themselves in stage four, other scrappers were quick to leave. To some degree, this stage and its length were dependent on the hour of the day. If a scrapper had found enough metal requiring a journey to the scrap yard before the day's end, they were often quicker to leave to locate more metal and return before the yard closed, whereas the later in the day scrappers arrived, the more likely they were to engage in conversation during stage four, as they did not plan to venture out again to search for metal.

These recycling center or yard relationships appear to be the basis for many of the friendships observed during the study. For example, when I asked scrappers about their relationships with other scrappers, many said something similar to Cowboy, "Oh yes! I've got several friends of mine out at the junk yard where I go at." The yard provides the physical location needed to meet other scrappers and grants crucial time during each stage of the sale process to develop relationships.

4.2.2   On the Streets

While the yard may function to kick off a relationship or strengthen a previously established friendship, it is evident that some relationships are built before contact at the yard. In other cases, relationships continued beyond the yard and into the search for metal. Either way, strong bonds seemed to be forged within the scrapping community. An extreme example would be the relationship Cowboy and his scrapping companions maintained. While I was discussing scrapping relationships with Cowboy, he said, "As a matter of fact, [I scrap with my wife and] my wife's ex-husband." Apparently, at some point in the scrapping team's history, Cowboy's wife was married to his scrapping partner, whom she divorced to marry Cowboy. Despite this spousal exchange, the group continues to maintain a healthy relationship, centered on the pursuit of metal. While this type of relationship may be inconceivable for many in general society, as well as for most scrappers, Cowboy indicated that as long as everyone was focused on metal, it was not an issue.

Cowboy's was not the only familial relationship encountered. Sean and Travis are a father and son team, Mrs. Williams scraps with her daughter, Cody and Amber are dating, and Gary and Frank are brothers. In fact, Frank helps Gary push his metal collected in old grocery carts down to the yard at the end of each day and uses his personal identification card to make the transaction because Gary does not have an ID (which is required by law). During an interview, Frank made the following comment:

> *Frank*: I just help him bring it down here. But, I don't really do the scrapping. He does most of it.
> *Interviewer*: So he finds it; you just help him bring it down here? That is awfully nice of you for a brother.
> *Frank*: Ha. It is! (chuckling)
> *Interviewer*: Sometimes it ain't worth it, I bet.
> *Frank*: No! Especially with the attitude he has. But I just look over it. Go on and do what I have to do and let it be.

Frank is not the only scrapper who helps the family. Samantha, a professional who scraps in the course of her business, contributed to establishing her cousin as a Professional Scrapper several decades ago:

> *Samantha*: I had a cousin that bought him a truck scrapping. He didn't have no education and you know, no high school education or nothing. And I got him into scrapping. He bought him a little truck scrapping, and then he started riding around, and that is all he does. He still does it.
> *Interviewer*: So you taught him how to scrap, and he has been doing it for a couple of years?
> *Samantha*: Well, he has been doing it since '86. I showed him in '86 how to scrap. He used to live, he was in the dumpsters, anyway getting cans and stuff. But, you know I showed him how to get junk. He would ride around with me, and we would go junking.

Even scrappers who were not related discussed their friendship when they crossed paths searching for metal. When I asked scrappers, "Do you get along with everyone else who's a scrapper?" the majority responded something similar to Jen, saying, "Yes, I do!" or like Gary, who eagerly added, "I get along with everybody." Frank stated that he had "at least four other [friends] who scrap. I get along with them very well. There are a whole lot of people who scrap."

While only some examples are included, there were many cases of the friendships and pleasant acquaintances established during scrapping. It is probable that scrappers are around one another frequently and thus likely to develop relationships. What is more, the negative stereotypes which general society may apply to those who belong to a subculture that involves rifling through what many in society may consider waste serves to forge stronger bonds between scrappers than with other members of society.

4.2.3   When in Need

The strong bond that scrappers share is further demonstrated by the frequency and diverse ways in which scrappers assist each other. This type of collaboration occurs at many levels including, but not limited to, locating metal, acquiring metal, and providing tools or instruction on how to separate metals. For example, Samantha discussed some elderly or disabled scrappers she knows:

> There are a couple of them that call me now, because they are like, "I cannot do this no more, I can't scrap no more," their back is bad and, "I can't do it, so, if I find anything I will call you, let you know and you can come and get it," you know. I said, "OK, keep me in mind." And if I can't, I know someone else that will go get it, like [my cousin], I will call him first.

In fact, this is one of the most frequent partnerships, when a Subsistence Scrapper needs a vehicle to transport metal. Often another scrapping friend or acquaintance who has a vehicle is called for assistance. Frank explained how this works for him:

> *Frank*: Right now, I have a bike. My plan is to have a Ford Fusion. But I do have [scrapping] friends that have a truck, so whenever I scrap and I got a pile, I call, and I am like, "Hey, I need you to come pick this pile up for me real quick. "
> *Interviewer*: What do they charge you for that?
> *Frank*: Just a few bucks for gas, 'cause they are friends.

By the same token, Brett, a Philanthropic Scrapper, talked about his acquaintances, who are Subsistence Scrappers. He said, "There are a couple of times he [a scrapping friend] has big loads, and I take him over there [the recycling center] 'cause he can't drive, so I take him." Similarly, Samantha said she had a scrapping friend, describing him as, "Friend of mine that is redoing a house and stuff and I am helping him take stuff out. I got a truck, and he don't, so he gives me a little bit of something [when I help]." Money is often exchanged for assistance involving a vehicle; typically, however, the money was insignificant and primarily used to cover car expenses.

Scrapper Profile

Name: Brett
Taxonomy: Philanthropic Scrapper
Location: Southwest—Inner City
Age: Mid-30s
Race: White
Education: Masters
Employment: Full Time—Public School Teacher
Experience: While observing activities at a scrap yard, I noticed a man walk in wearing a tie and a three-piece suit and holding a small motor. He explained that he did not scrap often, but always keep metal for several Subsistence Scrappers he knows, such as Craig, with whom I had spoken. However, he had not seen Craig for several weeks and wanted to dispose of this broken part "in an environmentally responsible way." In addition to collecting metal for scrappers, he would also drive them to scrap yards and use his ID to complete the transactions in an effort to help them. We discussed how many individuals treat scrapping as their job, and he said, "It helps make the world a better place, scrapping, I mean it's not work with dignity in the same way that other work can [be], but you don't have to be parasitic." As we continue to talk, the staff hand him is money for the small motor, and he pauses and looks at it, saying, "I thought I would get more than 44 cents."

We both laugh, and I ask where he lives. He explains that he chose to live in the inner city, "To keep my eyes open to the world, 'cause this kinda stuff gets forgotten, and somebody needs to pay attention. It's not that I feel like I am here to serve anybody, as much as I want to be a good neighbor. Yeah, just to learn how people live, 'cause I grew up in the south end, but it was middle class. Everybody in my neighborhood was generally blue-collar, they worked at [a local factory]. So it wasn't until I was in college that I had more than a two-word conversation with somebody who was homeless or knew much about addiction or scrapping or anything like that. If we are going to have a world of peace and justice, we need to know each other. So I thought I will live down here and work out there and I will have a foot in both worlds and feel like I am isolating myself in the community. I mean Craig, I bet he does not go outside of a couple mile radius, and I bet a lot of people who live down there are like that, but the same is true for the East End, separate ghettos, rich ghettos stuck in their own world."

Scrappers also tended to assist each other when they were having an unfortunate day at scrapping. Spoons mentioned, "The days I can't find nothing other scrappers come by and help me ...'cause they know, they got their truckload down, and they know I ain't got nothing." Likewise, Chad shared the following:

Some days it nasty weather, and some days it's the nicest days on earth, and I work my ass off and ain't finding much, and someone (indicating a fellow scrapper) will say, "Hey, you want some cans," and I will say, "Yes, please," you know.

David told me, "You know when times are rough, we try to find scrap [together]. Everybody is doing it," referring to the occasional and impromptu partnership he has with scrappers when he is struggling to make enough money.

Samantha who is a scrapper and maintenance worker for a local government apartment complex described how she proposed and instituted a division-wide policy designating a location where maintenance workers can leave unwanted metal parts and appliances to help those who are trying to make a living off scrapping:

> Well, we have a [concrete] pad where I work and whatever the contractors we have that come in and tear stuff out leave, [we put] the scrap and stuff on the pad. The pad is open for these guys that ride around in trucks and stuff, and it keeps us from having to get rid of it. So, we call them to come and get it. And, they can make them a little money or whatever.

Despite the competition to locate and secure metal as a means of income, scrappers are frequently willing to place helping each other above their pursuits. Whether that assistance is driving another scrapper and their metal to the yard, or giving a down-on-his-luck scrapper a little metal, scrappers demonstrate a strong bond and commitment to help each other. As Dustin put it, "We made sure each other got what they needed."

4.3   NO HONOR AMONG THIEVES

As has been identified, scrappers tend to function as a subculture sharing similar actions, behavioral norms, values, and codes. These common subcultural attributes appeared robustly across all locations and in most individuals who participated in the present study. However, thieves are an exception to the scrapping subculture. Practically all of those who were involved in stealing metal did not act or think like most other scrappers. Nor was there a cohesive concept of values, norms, and codes presented by thieves. Therefore, while nearly all thieves called their activities scrapping and many self-identified as scrappers, their thoughts, behaviors, and values were clearly not aligned with the majority of scrappers in the other taxonomies, nor did the taxonomy have clearly defined values and beliefs.

While, in general, thieves and scrappers had different values, norms, and codes, there were a few similarities. The most common aspect of the scrapping subculture, among most thieves, was how thieves got their start in metal. Most explained that they began to scrap legally or as Dustin said, "We did it for about a year legally."

A few other thieves admitted that they often scrapped legally, but admitted that when they saw something they wanted, they would come back and steal it. David, who has an extensive network of metal theft friends, explained it by saying, "A lot of old boys do it legal, you know, but if it's something that they can't just take (implying stealing) they'll be back. They'll group up, they'll put a plan together, and they'll be back." Moreover, John provided the clearest example when he related the story of how he and his scrapping partner searched for metal on his family farm, but then drifted into theft:

> We [were] just [on] the family [farm], was hauling scrap one day, we rode past that field (neighbors), and it had a bunch of tobacco frames in it. Next thing you know, I don't know really which one thought it up. We thought about it, and we had just kind of ran out of scrap one day, and [when we did scrap] were coming out with about $20 [each]. So we just kind of pulled in over there with the truck, and just kind hooked up to the tobacco frame, and took it over there and got like $80 something, and I was like dude! Then we got to the point we was hauling one or two a day.

Another common difference between the subculture of scrappers and metal thieves were their beliefs, norms, and codes relating to territorial disputes. While scrappers tended to *show respect* to each other and operate under a code of *first come, first serve*, metal thieves were territorial, with no established code of conduct. For example, Dustin described several instances where there were conflicts between other metal thieves who were also stripping abandoned homes which he had "already spoken for," whereby he considered the territory claimed. He went on to explain, "If someone else was getting our metals, there were a couple of fist fights. ... [we] just pretty [much] told them to fuck off."

Dustin went on to explain, "There were some [thieves] you just hated because they would steal your stuff." He was not the only thief who mentioned this. Several thieves discussed how it was common to steal from another thief. Taking another thief's metal could occur at the thief's home, like at David's house, "While I was in jail they cut all of the wirings out of my shop and took my pump off the well ... all of my brass fittings off of the faucets ... just about everything that wasn't nailed down." Stealing another thief's metal even occurs at the original theft place. Leo, a thief who conducted highly organized repeated thefts of copper from abandoned buildings, discussed how other thieves would steal the material he was taking:

> *Leo*: There was one time I actually was going in there, and I'm walking up the hill, and I see these two guys, and when they see us they ran off into the woods. So I just figured they were just there doing something [that]

they weren't supposed to do, so I ended up going into the building. [Later in the day] I went outside, and one of the guys was running up the hill, and I told him, "Man, listen you don't have to, I'm not a cop. You don't have to run, whatever you want to do, go ahead and do it. There's plenty for all of us." I had about $2000's worth of copper on the floor ready to go, and [the guy] said, "You guys can't be here scrapping, my uncle is the caretaker." So at that point, you get scared, and you just pack up your stuff and leave. Well, we came back the next night, and all of our stuff was gone. They pretty much just said that to get what we already took.

Interviewer: So they got one over on you?

Leo: Yeah they got one over on us ... and that's happened a couple of times where we left stuff in the building and went to pick it up that night, and the stuff was gone.

Unlike scrappers, who maintained a level of trust with each other, most metal thieves did not trust each other. Jennings shared, "I've trusted people in the past, and they've screwed me." Chris explained that working with other thieves was troublesome because, "They are like, 'Oh, I helped you, I helped pull this out, and I helped do this, where is my half at?'" Moreover, thieves were even more unlikely to share territory than scrappers were. Jessica revealed, "You did not discuss what you done 'cause you didn't want them to know where you done went."

Another striking difference between metal thieves and scrappers is the company they keep. While most scrappers appeared legitimately hard-pressed to identify known metal thieves, the metal thieves seemed to know each other. For example, David said of his acquaintances who are metal thieves, "A dozen would be a light number, there's probably 12 to 20, and there's not a one of them that won't steal." Similarly, Dustin estimated he knows, "20 people legit and about 15–20 that stole."

Similar to the scrapping subculture, some of these theft acquaintances or friendships were made at the scrap yards. Dustin explained how he met other thieves:

Dustin: At the scrap yard or going into properties

Interviewer: You talked about meeting other scrappers that stole at scrap yards, how did you know them as stealing or did you just start talking and find out later?

Dustin: Most found out later, but when you see certain things come in you know they are stolen.

As Dustin mentioned, some relationships with other thieves were forged when meeting while stealing or burglarizing a building and "trying to get the same stuff." However, a few other thieves mentioned "meet[ing] through the drug world" or other criminal relationships.

Regardless of how thieves met, they were unlikely to help each other unless they were in an established partnership. Chris talked about working with other criminals:

[I] just prefer to do it alone. I ain't trying to split something with nobody. It's not that I am fucking being a dick. I'm not being greedy. I don't have nothing. So the more of what I can get the better.

Chris's comments were relatively common across thieves. There was a general lack of trust of one another. For example, Michael described his relationships with his partner, saying, "Who knows, he might not have even given back equal splits because the yard we used didn't usually even give a receipt." Finally, the few times any assistance was given to another thief it was only to "the ones that helped us out as well."

4.4   Summary

This chapter has revealed that scrappers have their own subcultural norms, values, and mores. These behaviors and beliefs may be described as a *scrapper's code* and include concepts to reduce territorial conflicts such as *first come first serve* and *show respect*. In addition, to avoid conflict, scrappers had a near universal disdain for metal thieves, proudly claiming they were "not a thief." Not only did most scrappers not associate with metal thieves, they had a well-formed understanding of what was available for taking and what was not. Developing a code of *ask first* whenever doubt existed of the availability of an item was to keep scrappers from stealing or being accused of a crime.

Moreover, scrappers tended to be part of a community of other scrappers and had many social and entrepreneurial connections with one another. These relationships were forged by encounters at the recycling yard, during the search for metal, and due to familial connections and relationships. These relationships generally maintained strength, and when one scrapper was in need of help, either with a large load or due to a lack of funds, other scrappers tended to offer assistance.

However, most of these values, beliefs, and behaviors are largely absent from thieves, and in a most instances appeared contradictory to the scrappers' code. Overall, metal thieves were territorial, suspicious of each other, tended not assist one another, and often knew many other thieves. The majority of metal thieves were not simply scrappers who stole, but rather functioned separately and outside of the subculture of scrappers.

REFERENCES

Miller, J. M., Schreck, C. J., & Tewksbury, R. (2011). *Criminological theory: A brief introduction*. Pearson Higher Ed.

Primer on Metal Theft

History, Harms, and Prevalence of Metal Theft

At this point in the book, the attention is exclusively focused on metal theft. In particular, this chapter examines the types of metal commonly stolen, by providing a primer on metals (e.g., precious, ferrous, and non-ferrous). Next, a historical account is pieced together to help understand why metal theft became such an epidemic in the United States. After gaining an understanding of what metals are commonly stolen and acquiring a historical understanding of what triggered the spike in metal theft, this chapter examines the prevalence of metal theft. Finally, it explores some of the costs, both direct and indirect, to society resulting from metal theft.

5.1   Primer on Metals

The three broad types of metals are precious, ferrous, and non-ferrous. Precious metals include gold, silver, platinum, rhodium, and palladium. These metals have been historically valued and have had significant uses for thousands of years. Precious metal crimes often take the form of jewelry theft. The stolen jewelry is either kept for personal use or sold to jewelry purveyors and is rarely altered from its original form (e.g., melted to form other jewelry). While jewelry theft is common (FBI, 2014; Jewelers' Security Alliance, 2014), it is not the focus of the present study. The only exception is a discussion of precious metals stolen as part of other components (e.g., platinum in catalytic converters) and sold to recycling centers where their form is altered.

© The Author(s) 2017
B.F. Stickle, *Metal Scrappers and Thieves*,
DOI 10.1007/978-3-319-57502-5_5

The second type of metal frequently stolen is ferrous metals. Ferrous refers to the presence of iron in metal, which results, among other things, in ferrous metals being magnetic. Ferrous metals are commonly used in society in the form of automobiles, appliances, cans, structural steel, stainless steel, and more. Further, ferrous metals are the most commonly recycled metal in the world (John Dunham and Associates, 2013). Given the weight and size of many ferrous items, however, theft is not as common as with non-ferrous metals.

The last type of metal discussed in the present study is non-ferrous metals such as copper, aluminum, nickel, tin, zinc, and lead. As these metals do not contain iron, they are non-ferrous. These metals are often referred to as base metals since they are combined in varying degrees to create different alloys such as brass. Non-ferrous metals are generally more valuable than ferrous metals due to their properties, uses, and limited availability in nature. Non-ferrous metals are low weight, non-magnetic, resistant to corrosion, and are excellent conductors of electricity. Base metals are preferred for theft due to the low weight, smaller size, easy identification, and high value. Moreover, the frequency with which these base metals are stolen has led some to refer to them as semi-precious metals in an attempt to compare the value of non-ferrous metals in the current commodities market with the values of precious metals.

Metal recycling is the process of repurposing various metals into new uses. In most circumstances, this involves melting metals and reusing the metal to create new products. Recycling is an important part of the metal industry. Metal recycling reduces the expenses of mining new raw sources and is a vital and necessary component in the creation of some alloys.

5.2   AUTOPSY OF AN EPIDEMIC: WHY METAL THEFT RATES INCREASED

Historically, demand for metals is nothing new, nor is the concept that persons would acquire such metal illegally (Posick, Rocque, Whiteacre, & Mazeika, 2012). The United Kingdom, for example, had laws punishing the theft of metal from buildings and ships, as well as a history of court cases involving such thefts stretching back to 1674, when court record keeping began (Bennett, 2008a). Research, which used an analysis of newspaper reports, found that metal theft occurred in the United States on a regular basis around the turn of the twenty-first century (Whiteacre et al., 2012). There is even evidence to suggest that the nature and rate of metal theft in society is cyclical, corresponding with increases and decreases in demand (Bennett, 2008b; Kooi, 2010; Roggio, 1998; Sidebottom, Ashby, & Johnson, 2014; Sidebottom, Belur, Bowers, Tompson, & Johnson, 2011). If these analyses are correct, the United States is experiencing a significant increase in metal theft, due to demand cycles.

The current cycle began as copper and other metals were slowly rising in price in June 2003, after a historic low of 65 cents per pound on the London Metals Exchange in 2002. The growth occurred for many reasons, including an increase in construction in North America and Europe, increased raw materials needed for the US war in Iraq, commodities speculation, and, most important, an emerging Asia market—primarily China and India—which were experiencing substantial economic development and industrialization (Bennett, 2008a). Along with copper, the other three key non-ferrous base metals—aluminum, nickel, and zinc—became highly sought after, which caused their value to increase. On October 9, 2003, the Grasberg copper mine in Indonesia—the second largest supplier of copper in the world (International Copper Study Group, 2014)—suffered a significant mine collapse and slowed production. The decreasing supply and increasing demand for copper and other metals captured the attention of commodities speculators. They began to buy and trade metal futures, profiting by millions of dollars. The result of this speculation, however, was a disconnection between realities (supply and demand) and pricing of raw materials (Bennett, 2008a). Speculation led, in part, to an artificial increase in the price of copper, as well as other base metals.

The commodities market was stimulated even further in 2006 when the US Geological Survey announced that more than 25% of the world's copper ore had already been used (Papp et al., 2007). At the same time, Lester Brown (2006) claimed that within 25 years the supply of raw copper ore would be depleted. These theories caused some to believe that peak copper would soon be reached. Peak copper refers to the theoretical point at which maximum global production of copper is achieved, while demand continues to rise, resulting in a limited supply of copper—a finite resource—and thus increased prices (Simon, 1998).

These culminating events triggered a sharp price increase, which peaked at over $4 per pound for refined (non-scrap) copper by mid-2006. The increase represented a 515% increase in the price per pound of copper in less than a decade. As the price continued to rise, commodities speculators continued to speculate, and China and other countries continued to buy. In fact, global consumption of copper increased 41%

between 1996 and 2006 (Jolly, 2009), yet copper mine capacity only increased 3.7% during the same time (International Copper Study Group, 2014). Due to the low price of these metals in the 1990s, many mines and refineries failed to increase or update mining and smelting capacity. The result was a global deficit, specifically in copper, where consumption outpaced production. As production slowly began to rise, recycling filled the gap. Recycled copper accounted for only 13% of refined copper in 2005. However, it gradually increased to 18% of all refined copper by 2012 (International Copper Study Group, 2014). As 2006 rolled into 2007, the demand for copper and other metal prices began to wane along with the slowing of the global economy. Near the beginning of 2007, copper prices had settled around $3 a pound and fluctuated significantly between $2 in 2009 to $4.58 in 2011.

As copper and other base metals reached historically lucrative prices in 2006, many individuals began to cash in by recycling metals. It seemed that everyone began to pay attention, from the farmer cleaning out his junk pile, to the plumber collecting scrap metal after a job, to the businessman stopping to pilfer through the trash set out for pickup at the curb, to the less scrupulous who would steal it anyway they could. When the price was right, people took notice that metal was everywhere and ripe for the taking. About the same time, reports of metal theft began appearing all across the United States. Items such as bronze plaques, catalytic converters from vehicles, brass fittings, aluminum siding, grounding wire, copper piping, electrical cables, air conditioners, and even the brass nozzles off fire truck hoses began to disappear at an alarming rate. Metal theft emerged as one of the fastest growing crime trends in the United Kingdom (Bennett, 2008a) and the United States (Kooi, 2010; Whiteacre, Medler, Rhoton, & Howes, 2008).

While other metals were included in this rush to cash in on scrap metal recycling, copper became the most sought after metal, leading some to refer to the increase in demand as a "red gold rush" (Berinato, 2007; Bobrowich, 2013; Smith & Craze, 2010). This high demand for copper is due to several of its unique characteristics. There are few substitutes for copper's many applications—from communication and electronics components, to transportation and construction. Copper is the most widely used metal in the world (Copper Development Association, 2014) due to its ability to resist corrosion, conduct heat and electricity, and its antimicrobial qualities (International Copper Study Group, 2014). Due to its extensive usage, copper prices are often used as a proxy for base metal prices (Posick et al., 2012). However, copper is not the only metal frequently recycled. For example, recycled ferrous metal (steel) serves as a vital chemical component when manufacturing new ferrous metals and cannot be substituted by virgin iron ore (Bennett, 2008a). In other words, recycled metals are necessary for the creation of new non-recycled metals.

During times of high demand, the value of recycled (used) copper can be as high as 90% to 95% of the price of refined raw copper ore (Copper Development Association, 2014; Davies, 2008). Moreover, secondary copper (recycled or scrap copper) is appealing when compared with many other metals, because the recycling process does not negatively affect the chemical and physical properties of copper (International Copper Study Group, 2014). Due to this recyclability and the high demand and sluggish production of copper ore, by 2011 approximately 35% of all copper consumed in the United States was sourced from recycled copper (International Copper Study Group, 2014).

The historical demand for all types of metals has soared in recent years. Driven by increased uses in electronics, construction, and market speculation, the production of metal has not kept up with demand. This has served to increase the price of recycled metals, as the manufacturing industry has turned to scrap metal to meet the demand. While there is evidence that metal prices are cyclical, consumption does not appear to be waning globally. Bennett (2008a) concluded that "[h]igh prices, readily accessible materials in the built environment and informal infrastructure for the sale and integration of stolen metals back into the production chain create a cycle of asset stripping that has major economic and social costs" (p. 182).

5.3   PREVALENCE OF METAL THEFT

With the unprecedented increase in demand coinciding with a dramatic price rise for secondary metals, metal thefts began to grow. At least that is what news reports and government agencies indicated. It seemed that news reports of extraordinary metal thefts filled the pages of local papers and made their way onto local and national television shows, as well as internet sites. Industries and government agencies began to call the increase in metal theft a serious problem (Institute of Scrap Recycling Industries, Inc., 2011) and a threat to the critical infrastructure of the United States (FBI, 2008).

Unfortunately, a significant vacuum of knowledge existed. No one was sure of the exact amount of metal stolen in the United States, who was stealing it, and the exact cost to society. Moreover, law enforcement agencies typically did not separate metal theft data from other types of theft. This lack of reliable information led many to rely on news media reports and anecdotal evidence as a measure of metal theft and its effects (Kooi, 2010). While there is little doubt that metal theft has increased, based on these reports there is insufficient empirical research to determine generalizable causes, or even the frequency, of metal theft (Bennett, 2008a; Kooi, 2010).

5.3.1   Council of State Governments

Dissatisfied with the void of data on metal theft and knowing the significant costs to state governments associated with these thefts, the Council of State Governments (CSG) launched a national investigation into metal theft in 2014. The results were discouraging, finding there was no comprehensive national source of metal theft data. CSG next turned to each state, and yet again, found no state retained comprehensive data on metal theft. Finally, CSG interviewed more than 200 law enforcement representatives. The results were mixed, with a few jurisdictions keeping records, but the techniques and types of data varied widely, as did the methods for retrieving the records (many required a keyword search, which CSG found to be unreliable). This led the CSG researchers to conclude that comprehensive empirical metal theft data at national, state, and local levels were not available (Burnett, Kussainov, & Hull, 2014).

5.3.2   National Insurance Crime Bureau

Despite the fact that no national or state governments keep data on metal theft, a private organization does track insurance claims of metal theft in the United States. This data, which is the most reliable and largest source of metal theft data currently published, is collected and analyzed by the National Insurance Crime Bureau (NICB). The data the NICB utilizes is extracted from the Insurance Services Office (ISO) ClaimSearch, which is a clearinghouse where insurance companies submit claims data. Unfortunately, the data contained within the NICB's reports only represent insurance claims, which has three significant flaws. First, metal theft data may be missing due to the insufficient monetary loss necessary to file a claim, or the property may not have been insured and thus no claim filed (Burnett et al., 2014). Second, the data were gathered through a query from the ISO ClaimSearch database for keywords such as "theft," "took," "steal," and "missing" combined with the terms "copper," "bronze," "brass," and "aluminum." Therefore, significant areas of metal theft may not have been included. For example, Whiteacre, Terheide, and Biggs (2014) found that 25% of all metal thefts in Indianapolis listed in police reports were appliances, and it is doubtful that these were included in the ISO keyword search results. Moreover, misspellings of different types of base metals or the listing of other types of metal not included in the search (e.g., steel, iron, lead) would not be captured. Finally, the third significant flaw, according to NICB, is that "the average delay between the theft occurring and the claim record entering ISO ClaimSearch was 31.3 days. [Therefore,] some thefts that occurred within the later months of [the] analysis may not have been entered at the time the data was collated" (Kudla, 2009, p. 4), resulting in missing data. The outcome of the NICB efforts is likely a significant underreporting of metal theft rates across the United States. Regardless of the potential shortcomings, the NICB reports provide the only national level data available on metal theft. Moreover, the findings are frequently utilized as the proxy for the rate of metal theft across the Unites States. Thus, it is important to examine these reports.

The first widely available report on metal theft by the NICB took the form of an analysis of catalytic converter thefts. A catalytic converter is part of a motor vehicle's exhaust system, which reduces the toxicity of emissions. The unit is mounted in-line with the exhaust system under a car and can be easily and quickly removed with a metal saw. The catalytic converter contains a small amount (3 to 7 grams) of the platinum metal group such as rhodium, platinum, and palladium (Specialty Metals Smelters & Refiners, 2014). At the time of the NICB's research, catalytic converters were selling at recycling centers for between $20 and $200 per unit. NICB queried ISO's ClaimSearch database and discovered 1388 claims from the first six months of 2008, representing a dramatic 3000% increase over the prior year (Stanfill, 2008). No other statistical analysis was performed, and the report stated, "The NICB cannot, at this time, accurately calculate the total number of catalytic converter thefts nationwide. There is, however, a significant amount of anecdotal data such as media reports, interviews, claims information and NICB cases that indicate catalytic converter thefts are a growing and widespread problem" (Stanfill, 2008, p. 4). NICB has not published a follow-up study examining catalytic converter theft, and no other organization is known to have studied the topic. Moreover, no other organization or government entity is known to track thefts of catalytic converters. Thus the current trends and prevalence remain unclear. The second report released by the NICB in 2009 again examined insurance claims utilizing ISO's ClaimSearch. However, this search excluded catalytic converters and instead considered base metals (copper, aluminum, brass, and bronze), as well as precious metals (gold and silver). NICB collected theft data over 35 months between 2006 and 2008. The results indicated that there were 13,861 claims for base metals (Kudla, 2009), or an average of 396 per month. The report also found that 94% of the claims listed copper as the metal stolen, that approximately two out of three thefts occurred on commercial policies, and that 62% of all claims were for utility, construction, or housing components such as piping, wire, plumbing, siding, cable, and the like. There were 2376 claims on precious metals, with 90% occurring on personal policies. NICB provided a listing of the top 10 states and cities with the highest claims of both base and precious metals; however, the rankings were not weighted according to population and are of limited value. NICB concluded the report by indicating that base metal theft accounted for nearly six times the thefts of precious metals, and discussed the relationships of theft with the high prices of metal and the easy access to base metals as driving the trend.

The NICB released its third report relating to metal theft in 2012. The report examined the insurance claims of base metal thefts reported from January 2009 through December 2011. The investigation revealed 25,083 claims submitted for metal theft (Kudla, 2012), indicating an 81% increase between 2009 and 2011, of which copper accounted for 96% of all claims. The report continued by examining the relationship between price and copper theft as well as claiming a relationship between drug usage and copper theft. The report also identified that 55% of the claims were from commercial policies, with the remainder on personal policies. This time NICB evaluated the rates of metal theft claims per 10,000 residents in each state. Findings indicated that the top five states for claims of metal theft were Rhode Island (2.587), Delaware (2.039), Ohio (2.077), Kentucky (1.781), and Georgia (2.039) (Kudla, 2012). Unfortunately, very little discussion or statistical analysis is provided in this report, and it is difficult to draw conclusions other than a snapshot of national insurance claims for metal theft between 2009 and 2011.

The NICB followed up the 2012 report with a very brief media announcement in 2014 stating that insurance claims of base metal theft between 2011 and 2013 dropped by 26%. However, this analysis did not include a monthly statistical categorization as did the previous reports. Thus a monthly comparison of metal theft trends is not possible. A request for month-to-month data for the present study was denied. The 2014 NICB media release identified just over 41,000 claims related to copper, bronze, brass, or aluminum between 2011 and 2013 revealing 97% of all claims were for copper theft. The report concludes by stating, "NICB sees hopeful evidence that the national problem of metal theft might be decreasing" (NICB, 2014, p. 1). NICB bases these hopes on legislative efforts aimed at crime prevention, police response, and citizen awareness; however, it provides no empirical support for these claims.

The data collected by the NICB has substantial design flaws and is subject to evaluating only the insurance claims for metal theft. Moreover, the data collected has not been distributed in a method that allows for advanced statistical analysis. Despite the flaws, however, it is the only data available, to date, on the long-term national trends of metal theft. What is more, it was the first organization to evaluate the relationship between prices, drug usage, and other variables along with metal theft. Despite its weaknesses, it is an important benchmark for future studies.

5.3.3   Indianapolis Metal Theft Project

With the lack of national and state data on metal theft, the only other resource available to determine the prevalence of metal theft is at the local level. While most localities do not track metal theft (Burnett et al., 2014), Indianapolis, Indiana, does. This data collection effort was spearheaded by Dr. Kevin Whiteacre of the University of Indianapolis Community Research Center (CRC). In collaboration with the Indianapolis Metro Police Department (IMPD), the CRC provided the first empirical study conducted in the United States on many aspects of metal theft. Whiteacre identified a lack of scholarly research in the area of metal theft, also observing that most organizations, governments, and individuals rely on anecdotal experiences or news stories, which frequently portray "fantastic crimes" (Whiteacre et al., 2008, p. 6). IMPD and CRC worked in collaboration to "gather and analyze a wide variety of data that will provide a clearer understanding of the incidence, types, costs and impacts of metals theft in Indianapolis in order to inform and implement strategies to reduce these crimes" (Whiteacre et al., 2008, p. 4).

The first effort toward this goal was accomplished through a pilot study by collecting and analyzing data from IMPD crime reports between

January and March 2008. Researchers identified 678 reported instances of metal theft in Indianapolis, which converts into 7.7 metal thefts per 10,000 persons living in Indianapolis. Personal residences accounted for more than half of all reports (55%) with 24% occurring at commercial businesses, 15% automobile related, and 5% taking place at churches. Copper was the most commonly stolen item accounting for 32% of all metals. Researchers also examined the cost of metal theft, finding an average damage estimate of $4314 per theft incident.

The most recent publication from the Metal Theft Project is a Research Brief published by the CRC examining data covering a 24-month period (October 2011–September 2013). This report represents the first detailed time series analysis conducted in the United States on metal theft. Whiteacre et al. (2014) found a considerable increase in metal theft from the 2008 pilot study, an increase of some 57% (an average of 11 metal thefts compared to seven per day in 2008). The study also discovered the theft of metals was involved in nearly 10% of all reported burglaries and thefts in Indianapolis in 2012.

Copper was again the most commonly stolen metal, accounting for 34% of metal thefts. However, appliance theft increased 127% from the 2008 study to account for 25% of all metal thefts in 2011 and 2012. Appliances often contain a small amount of base metals (e.g., copper) and are made structurally from ferrous metals (e.g., steel and tin). Whiteacre et al. concluded that appliance theft is an indication that metal theft is becoming "more organized and purposeful than the odd thief with a shopping cart or backpack" (2014, p. 2). While the study found a substantial increase in instances of metal theft, there appeared to be a decline, by nearly half, in victim estimates of property loss with an average of $2034 per incident. While the data collected and analyzed within these studies is important, it should be noted that these efforts, as well as statistical trends, represent only one community and may not be representative of the nation as a whole.

5.3.4   Rochester, New York

The only other city known to keep data on metal theft and to publish the data is Rochester, New York. Chad Posick (2008) with The Center for Public Safety Initiatives at the Rochester Institute of Technology made use of data provided by the Monroe Crime Analysis Center. The data examined the theft of copper from residential and public structures, which were referred to as copper burglaries. The working paper published provided few statistics, but discovered that copper burglaries in Rochester averaged 30 per month in 2007 and 2008. Of those copper burglaries, 80% occurred in a vacant property.

In 2012, Posick et al. conducted a second analysis of the Rochester data. This study examined copper burglary over a 27-month period, from 2008 to 2010. During this time, 10% of all burglaries in Rochester were related to copper theft. The authors examined the data through a spatial and price-theft analysis using variables such as time, entry methods, vacant buildings, and more. Other sections of the present study will examine these areas in more detail.

5.4   COSTS AND HARMS TO SOCIETY AND INDIVIDUALS

While the prevalence of metal theft incidents is largely unknown, another common method for examining crime, the estimation of costs and harms, may provide insight into the damages caused by metal theft. There are often several approaches for identifying harm when examining crime. Some of the more popular methods are an examination of direct costs of the items stolen and the indirect costs and harms related to theft. Direct costs are the losses that are directly linked to the value of replacing the item taken. An example would be the cost or value of copper wiring stolen from an abandoned property and the expense of replacing it. The indirect costs associated with stolen goods include the costs related to repairing the damage done by thieves to acquire the metal, as well as the expenses related to increased insurance, productive time lost due to the theft, and more. Both types of losses are necessary to evaluate and understand when studying crime. Moreover, costs and harms, mainly indirect, are necessary to consider, given the nature of the offense, which often leads to significant indirect costs to individuals and society. The following sections will examine what is known about the costs associated with metal theft and its impact, direct and indirect, on persons and society as a whole.

5.4.1   Direct Costs

Estimating direct costs to individuals due to metal theft is a difficult task for two reasons. First, there is a lack of data available to quantify the amount or value of stolen metal. Second, the damages caused to obtain metals (e.g., damage to a structure when removing copper pipes) are often not recorded.

Image 5.1 While this image is taken of an abandoned house under renovation, it illustrates the costs associated with property damage related to metal theft. Thieves often target homes like this and pull the copper tubing from the floor or walls, causing more damage than the value of the pipes themselves

These two factors present a significant hurdle to ascertaining the actual direct cost of metal theft. As discussed previously, several organizations within the United States collect data on the prevalence of metal theft: the city of Rochester, New York, the Indianapolis Metal Theft Project, the NICB, and to some degree the US Department of Energy (DOE). However, only two, the Indianapolis Metal Theft Project and the DOE make any effort to identify and quantify the costs associated with metal theft. Since most utility and transportation companies are publically owned, the majority of their costs will be examined later under the indirect costs section.

The Indianapolis Metal Theft Pilot Study by Whiteacre et al. (2008) examined, among other data, the claimed value of metal stolen as reported to the police by each victim. The pilot program, covering the first three months in 2008, identified 678 instances of metal theft. Of those, 25% (169) of the victims provided estimates for the value of metals stolen. The average estimated loss was $4314, with a median loss of $1500. For those thefts in which a loss estimate was not included, the researchers calculated a loss based upon the average of known losses. Their findings concluded that approximately $1 million in losses each month occurred in Indianapolis during the first quarter of 2008.

The Indianapolis Metal Theft Project provided a second estimate in 2014, which examined police reports of metal-related thefts during 24 months (October 2011 to September 2013), identifying 8149 incidents of metal theft. The second report does not specify the number of persons who provided police with an estimated value of loss as did the first report in 2008. Therefore, it is not clear how the results were analyzed. Nevertheless, Whiteacre et al. (2014) concluded that there was a $16.6-million-dollar loss in Indianapolis over 24 months (or $690,000 per month), due directly to metal theft. The findings indicated the average loss was $2034 with a median of $537.

Notably, in both studies the researchers identified the potential weaknesses of the data and their conclusions. In the pilot study, they expressed concern calculating average losses for the entire group based only on 25% of the respondents who provided an estimate. In both studies, the authors noted that the mean was substantially higher than the median, indicating that several high-value losses skewed the mean upward. Finally, in the 2013 study, researchers stated that it was not clear if the victims were "estimating replacement costs, collateral damage, or just the perceived value of the item at the time of the theft" (Whiteacre et al., 2014, p. 3). Despite the potential weakness of the data, these two studies are valuable in the literature, as they are the only known analyses of the direct cost of metal theft to individuals in the Unites States.

As discussed previously, costs to replace the metal stolen may be significantly less than the damages caused to structures or devices that contain the metal. For example, a stolen 12-inch section of copper pipe which had emerged from a poured concrete foundation cost significantly more to repair than the value of the copper. The time, equipment, and resources necessary to remove enough concrete to re-attach a copper pipe to the existing line and then replace the concrete are often not considered when discussing metal theft. Other examples include the damage done to a structure when water pipes are removed and flooding occurs, or when wires are torn from behind drywall. Additional costs include repairing devices or materials damaged by thieves to gain access to metal. These may include windows, doors, or the destruction of an air conditioner unit to obtain copper (Posick et al., 2012).

Each of these examples and many others demonstrate the vast extent to which damage and expenses may occur during metal theft. This extensive array of damages and the substantial costs associated with them are not typical with other types of theft or crime. Thieves may be stealing only a few dollars' worth of metal, but cause thousands of dollars in collateral damages to acquire it (Bennett, 2008a; Posick, 2008). A survey by the Electrical Safety Foundation International (ESFI) (2009) of electrical utility companies found that the average cost to repair a single instance of metal theft was more than two-thirds the value of the metal. In total, the survey indicated that utilities spent $20 million to replace the actual copper item stolen (often copper wire or transformers) and $60 million in related expenses to repair the item or facilities damaged and other costs.

Unfortunately, the direct cost to individual victims of metal theft is hard to quantify. Once again, the primary problem is a lack of data. The majority of what is known is based on estimates from individuals. With the exception of data from the Indianapolis Metal Theft Project, estimates are usually conveyed through news reports and other anecdotal stories and may or may not include replacement costs, collateral damage, and other relevant costs associated with metal theft. Despite the lack of data, it does appear that there are significant direct costs to individual victims. Direct costs of metal theft should not only include the item stolen, but the damage caused to complete the theft as well as the labor and other factors that are necessary to make the victim whole. Until more data is gained in this area, real direct costs of metal theft will remain unknown.

5.4.2   Indirect Costs

The victims of metal theft are arguably the entire population (Blythe, 2008). While the largest percentage of thefts occur on individual private property, the remainder of thefts take place at businesses (Wilkinson, 2012), government utilities (DOE, 2010), churches (Walter, 2011), foreclosed properties (Kooi, 2010), and even cemeteries (Breen, 2008). The result is a huge number of persons who are directly impacted. However, this list does not include individuals who experience the indirect costs and harms of metal theft, including those who suffer from higher insurance costs (Wilkinson, 2012), increased utility costs or service delays (DOE, 2010; ESFI, 2009), interruptions in travel (GLA Transport Team, 2012), missing road signs and manhole covers (Alusheff, 2012), and other nuisances and harms that arise due to metal thefts. What is more, the indirect costs of metal theft extend well beyond lost metal (direct costs) associated with metal theft (Posick et al., 2012). The following sections identify many of the indirect costs related to metal theft.

Utility companies are frequently the target of metal theft and the findings discussed here represent the direct costs to utility companies, but more important, represent the indirect costs and inconveniences to utility customers. Clearly, any costs created by downtime of the electrical grid are borne out by individual users in an indirect way. Those who are left without electrical service suffer from the inconvenience of not having power, heat, or other amenities. Many may also suffer financially if they operate a business that relies heavily upon electricity, such as a grocery store or manufacturing company. Beyond issues related to the inconveniences and potential lost revenue due to power interruptions, the cost to repair the utilities must also be considered. Many utility companies are publically owned or rely on service fees; therefore, it is likely that costs associated with metal theft will be passed along to utility customers.

Utility companies, which rely heavily on copper and other base metals to function, include electrical facilities, cellular towers, telephone communications, railroads, and water companies. These institutions are also prone to significant losses due to metal theft. Governmental agencies warn of devastating effects on utility companies if metal theft continues (DOE, 2007, 2010; FBI, 2008). In 2009, ESFI conducted a survey of public utilities in the United States and discovered 95% of utilities experienced incidents of copper theft in the last 12 months and 81% were extremely concerned about copper theft. According to ESFI, the annual loss due to copper theft nationally from utility companies was more than $80 million in damages and additional expenses.

Utilities are not the only public organizations that experience indirect costs associated with metal theft. Many cities have indicated that they suffer significant losses related to metal theft due to damaged or destroyed infrastructure. For example, the 2008 economic decline affected many homeowners, resulting in an unprecedented rate of foreclosures; many of these foreclosed houses were abandoned, and ownership assumed by the local or state governments. Abandoned homes present a significant issue, as they are prime targets for metal theft (Kooi, 2010; Posick, 2008). In fact, Posick's 2008 study of copper-related burglaries in Rochester, New York, found that nearly 80% of copper-related burglaries occurred in abandoned structures and the city owned many of those properties. Moreover, these thefts often left costly damages to the property, including flooded basements, gas leaks, wall and floor damage, and other costs related to thieves removing pipes and wires (Posick, 2008). While Posick's study did not specifically examine the costs to the city, these types of damages are costly and significantly affect the value of the property. This damage can substantially affect the owners, whether they are banks, individuals, or local governments, as expensive repairs must be made before resale.

Another example of damaged or destroyed property related to metal theft comes from the Ecclesiastical Insurance Group, an insurance company that insures more than 90% of church properties in the United Kingdom. Ecclesiastical published a report in 2014 indicating that between 2007 and 2013 it received 11,000 claims from churches covered by its insurance due to metal theft. The most commonly stolen items were lead and copper roof materials, followed by copper wires, gutters, statues and other metal objects, including bells. Ecclesiastical estimated the total financial loss at £28 million or just over US$42 million. Ecclesiastical characterizes the loss of valuable heritage to the structures (many of which had original lead roofs dating back over 200 years) as irreplaceable. The insurance group also discussed the related financial losses arising from the thefts as a major problem, which small congregations are unable to bear. These costs included "damage to the stonework caused during the theft ... water damage to interior furnishings" (p. 1) as well as other related damages. The loss of "irreplaceable heritage" (p. 4) and costly repairs to damaged structures indicate a significant indirect cost related to the theft.

Another example of the indirect costs associated with metal theft is the damage done to the British railway system. It is estimated that between 2008 and 2011 the rail system lost more than £42 million or around US$64 million in repairs and other expenses related to the theft of metal (NetworkRail, 2011). Beyond this financial loss, the theft of energized copper cable delayed trains by more than 365,000 minutes, affecting nearly four million passengers. While the costs of repairs clearly affect the operating expenses of railroad, the slow travel time may have had an indirect impact on those serviced by the railway line.

Unfortunately, these examples are the only known published data on the indirect costs related to metal theft. However, they clearly indicate a significant loss in both direct replacement and related expenses of metal theft to government and other private jointly owned properties. These thefts are likely to affect the financial status of these organizations and the individuals connected with them. Beyond mere financial costs, there are other intangible costs associated with metal theft. The loss of a historical roof, damage to a historical structure, the inconveniences of missing street signs, bleachers, or playground equipment, and the loss of productivity related to power interruptions and travel delays are indirect costs that are difficult to calculate, yet are experienced by many in a community.

One final important indirect cost to consider is the costs associated with efforts to prevent or mitigate the effects of metal theft. These efforts may include physical security measures, hiring security guards, and costs related to the investigation and prosecution of metal thieves. While there is no published information to determine the degree to which individuals and organizations engage in these practices, anecdotal reports and casual observations make it apparent these efforts are common. Whether due to being a former victim or out of fear of becoming a future target, it appears that many individuals and organizations are taking preventative measures in the hopes of negating costly losses and repairs due to metal theft. As one recycling center owner said, "If you don't want people to take the metal, you've got to start treating it like what it is—an asset" (Berinato, 2007, p. 3). It would appear that many people are treating metals as an asset, which affects the finances of most organizations and individuals.

Utility and transportation companies seem to be at the forefront of metal theft, both in victimization as well as in of efforts to prevent theft and mitigate damages. The 2008 survey by ESFI found that nearly 95% of electrical utility companies in the United States have changed storage or security procedures to prevent copper theft. Efforts included installing security cameras, increasing fencing, additional signage, installing alarm systems, altering copper material storage and handling, improving lighting, enhancing visibility, increased security patrols, and other techniques. Many of these changes in procedures and physical structures are costly. For example, the survey found that utilities spent a combined $26 million between 2003 and 2008 to prevent or mitigate the threat of copper theft. This amount is significant, as metal theft was not typically a concern until late 2006 or early 2007.

Furthermore, many individuals and businesses may experience indirect expenses and inconveniences due to prevention and mitigation efforts. These effects may manifest themselves in criminal justice costs, inconveniences of legally selling scrap, and increased legal requirements on business. Prosecution remains an important aspect of metal theft deterrence (Southwire Company, 2012). Effective prosecution, however, requires personnel and other resources from the law enforcement community, the legal system, and the corrections system. Nonetheless, many of these institutions are suffering from diminished budgets and high workloads. Daniel Waterfield, Assistant Counsel and Director of Government Relations at ISRI stated, "[Police] don't have the resources and manpower to deal with what's traditionally been considered a low-level property crime" (Waterfield, 2014, para. 5). The difficulty in investigating the cases (e.g., most stolen metal has no identifying marks and is quickly disposed of) and legislation which fails to provide adequate penalties for metal theft result in those who are convicted rarely serving prison or jail time (DOE, 2007). Low prosecution and sentencing rates may lead law enforcement to not "take the time to arrest if [prosecutors] won't do anything" (Holeywell, 2014, para 6). The difficulty and expenses involved in investigating metal theft cases, combined with limited legal remedies and lack of awareness of the prevalence, result in infrequent investigations and in even fewer convictions. The infrequency of investigation and prosecution may embolden metal thieves and thus indirectly encourage a "migration to [metal theft from another more] risky enterprise" (Southwire Company, 2012, p. 5).

5.5   Summary

The demand for all types of metals has soared in recent years, especially for copper and other non-ferrous metals. Unfortunately, the production of metal through mining has not kept up with industry's needs. The increasing demand, coupled with the lack of resources, has caused an increase in the price of recycled metals as the manufacturing industry has turned to scrap metal to meet the shortage. As a result, the price of used copper has skyrocketed over 500% in just a few years. Coinciding with a worldwide economic decline, many individuals saw an opportunity to make money by recycling metal.

With the price of used copper and other metals increasing, the rate of metal-related thefts also increased. Unfortunately, there is little empirical data on the prevalence of metal theft in the United States. The best estimates available come from national insurance claims and the efforts of two cities to track metal theft. The NICB found over 25,000 insurance claims for losses related to metal between 2009 and 2011 (Kudla, 2012). Moreover, Whiteacre et al. (2008) found that the rate of metal theft in Indianapolis, Indiana, was a startling 7.7 incidents per 100,000 residents. Finally, studies in Rochester, New York (Posick et al., 2012), and in Indianapolis, Indiana (Whiteacre et al., 2014), found metal theft was a factor in 10% of all burglaries reported to the police. These studies, while limited in scope, indicate that metal theft is a growing problem.

As has been demonstrated in this chapter, there are significant collateral harms and costs associated with metal theft, which extend well beyond the crime itself (Posick et al., 2012). These costs include not just the direct costs to replace the metal stolen, but also the indirect costs to repair damaged metal components, labor expenses, the inconvenience of utility outages, expenditure increases needed to implement crime prevention efforts, and more. These losses occur directly to both individuals and organizations, as well as indirectly to society. In fact, the indirect costs and harms associated with metal theft are likely more significant than many other types of theft.

REFERENCES

Alusheff, A. (2012, April). Illegal scrapping in Ohio highest in country, hurting businesses and residential areas [Web log post]. Retrieved from http://blogs. bgsu.edu/bgreports/2012/04/26/illegal-scrapping-in-ohio-highest-in-country-hurting-businesses-and-residential-areas/

Bennett, L. (2008a). Assets under attack: Metal theft, the built environment and the dark side of the global recycling market. *Environmental Law and Management, 20,* 176–183.

Bennett, L. (2008b). *Metal theft: Anatomy of a resource crime.* Unpublished. Retrieved from http://shura.shu.ac.uk/4125/

Berinato, S. (2007, February). Red gold rush: The copper theft epidemic. *Chief Security Officer.* Retrieved from http://www.csoonline.com/article/2121953/loss-prevention/red-gold-rush--the-copper-theft-epidemic.html?page=1

Blythe, K. (2008). *Report from the survey subcommittee on Oregon metal theft collation: Results of concurrent surveys of law enforcement, property crime victims, and scrap metal recyclers.* The Oregon Metal Theft Coalition—Survey Subcommittee.

Bobrowich, J. (2013, December). Copper theft: "The red gold rush". *Alberta Community Crime Prevention Association.* Retrieved from http://albertacrimeprevention.com/december-column-copper-theft-the-red-gold-rush/

Breen, T. (2008, July). Cemeteries have new problem: Metal thefts. *USA Today.* Retrieved from http://usatoday30.usatoday.com/news/nation/2008-07-16-graves_N.htm

Brown, L. R. (2006). *Plan B 2.0: Rescuing a planet under stress and a civilization in trouble.* New York: W. W. Norton & Company.

Burnett, J., Kussainov, N., & Hull, E. (2014). *Scrap metal thefts: Is legislation working for states?* Lexington, KY: The Council of State Governments.

Copper Development Association, Inc. (2014, June). Copper facts. Retrieved from http://www.copper.org/education/c-facts/

Davies, L. (2008). *The lure of copper.* Brussels: European Copper Institute.

Ecclesiastical Insurance Group. (2014). *Guidance notes: Theft of metal.* Gloucester, GL: Ecclesiastical Insurance Group.

Electrical Safety Foundation International. (2009). *Copper theft baseline survey of utilities.* Rosslyn, VA: Electrical Safety Foundation International.

FBI. (2008). *Copper thefts threaten U.S. critical infrastructure.* Washington, DC: FBI Criminal Intelligence Section.

FBI. (2014). *Jewelry and gem theft.* Retrieved from http://www.fbi.gov/about-us/investigate/vc_majorthefts/jag

GLA Transport Team. (2012, February). *Greater London Authority.* Retrieved from http://www.london.gov.uk/transport-blog/governments-announcement-metal-theft-step-right-direction

Holeywell, R. (2014, February). *Metal theft: A rarely-talked about problem with a big price.* Retrieved from http://www.governing.com/topics/public-justice-safety/gov-metal-theft-legislation.html

Institute of Scrap Recycling Industries, Inc. (2011). *Preventing metal theft.* Washington, DC: Institute of Scrap Recycling Industries, Inc.

International Copper Study Group. (2014). *The world copper factbook 2014.* Lisbon, Portugal: International Copper Study Group.

Jewelers' Security Alliance. (2014). *2013 annual crime report.* New York: Jewelers' Security Alliance.

John Dunham and Associates. (2013). *Economic impact study: U.S. Based scrap recycling industry (2013)—Executive summary.* Washington, DC: Institute of Scrap Recycling Industries, Inc.

Jolly, J. L. (2009). *The U.S. copper-based scrap industry and its by-products.* Copper Development Association.

Kooi, B. (2010). *Theft of scrap metal: Problem-oriented guides for police series. Guide No. 58.* US Department of Justice, Office of Community Oriented Policing Services.

Kudla, J. (2009). *Metal theft claims from January 2009 to November 2008*. Des Plaines, IL: National Insurance Crime Bureau.

Kudla, J. (2012). *Data analytics forcast report: Metal theft claims and questionable claims from January 1, 2009 to December 31, 2011*. Des Plaines, IL: National Insurance Crime Bureau.

National Insurance Crime Bureau. (2014). *NICB: Insured metal theft claims see three-year decline*. Des Plaines, IL: National Insurance Crime Bureau.

NetworkRail. (2011, November). *NetworkRail*. Retrieved from http://www.networkrailmediacentre.co.uk/News-Releases/RAIL-INDUSTRY-CALLS-FOR-TOUGH-NEW-MEASURES-ON-CABLE-THEFT-190c.aspx

Papp, J. F., Corathers, L. A., Edelstein, D. L., Fenton, M. D., Kuck, P. H., & Magyar, M. J. (2007). Cr, Cu, Mn, Mo, Ni, and steel commodity price influences, version 1.1. *US Geological Survey Open-File Report, 1257*, 50.

Posick, C. (2008). Examining the relationship between copper burglaries and the price of copper in Rochester, NY: Center for Public Safety Initiatives: Rochester Institute of Technology.

Posick, C., Rocque, M., Whiteacre, K., & Mazeika, D. (2012). Examining metal theft in context: An opportunity theory approach. *Justice Research and Policy, 14*(2), 79–102.

Roggio, A. (1998, October 14). Thieves find some scrap not worth stealing. *American Metal Market, 106.*

Sidebottom, A., Ashby, M., & Johnson, S. D. (2014). Copper cable theft revisiting the price–theft hypothesis. *Journal of Research in Crime and Delinquency, 51*(5), 684–700.

Sidebottom, A., Belur, J., Bowers, K., Tompson, L., & Johnson, S. D. (2011). Theft in price-volatile markets: On the relationship between copper price and copper theft. *Journal of Research in Crime and Delinquency, 48*(3), 396–418.

Simon, J. L. (1998). *The ultimate resource 2.* Princeton University Press.

Smith, E. B., & Craze, M. (2010, November). Copper rises 50% in "red gold" rush on China doubling usage. *Bloomberg News.* Retrieved from https://www.google.com/url?sa=t&rct=j&q=&esrc=s&source=web&cd=1&cad=rja&uact=8&ved=0CB8QFjAA&url=http%3A%2F%2Fwww.bloomberg.com%2Fnews%2Farticles%2F2010-11-03%2Fcopper-rises-50-in-red-gold-rush-on-belief-china-to-double consumption&ei=CbEdVa3OGsGuggS0o4CABQ&usg=AFQjCNGglcZ86bxHtLpE6RKfgdKMI-fNBg&sig2=y-UEGMmPxcrki4IHQj57Wg&bvm=bv.89744112,d.eXY

Southwire Company. (2012). Curbing the copper theft crime wave. Arlington, VA: Rural Electric Magazine. Retrieved from http://remagazine.coop/

Specialty Metals Smelters & Refiners. (2014, January). Retrieved from Specialty Metals: http://www.specialtymetals.com/faq/

Stanfill, J. (2008). *Catalytic converter thefts.* Des Plaines, IL: National Insurance Crime Bureau.

U.S. Department of Energy. (2007). *An assessment of copper wire thefts from electric utilities.* Washington, DC: Infrastructure Security and Energy Restoration: Office of Electricity Delivery and Energy Reliability.

U.S. Department of Energy. (2010). *An updated assessment of copper wire thefts from electric utilities.* Washington, DC: Infrastructure Security and Energy Restoration: Office of Electricity Delivery and Energy Reliability.

Walter, N. (2011, July). Why metal theft is good business (but not for you…). Retrieved from Ecclesiastical Insurance. http://www.churchbuild.co.uk/2011/07/metal-theft-church-buildings/

Waterfield, D. (2014, February). Metal theft: A rarely-talked about problem with a big price. (R. Hoeywell, Interviewer)

Whiteacre, K., Medler, L., Rhoton, D., & Howes, R. (2008). *Indianapolis metals theft project: Metal thefts database pilot study.* Indianapolis: University of Indianapolis, Community Research Center.

Whiteacre, K., Terheide, D., & Biggs, B. (2014). *Research brief: Metal thefts in Indianapolis October 1, 2011–September 30, 2013.* Indianapolis: University of Indianapolis Community Research Center.

Whiteacre, K., Xia, Y., Zhou, Q., Zhou, B., Xie, W., & Posick, C. (2012). *Copper theft in the media (turn of the century).* Poster presented at the meeting of American Society of Criminology, Chicago.

Wilkinson, C. (2012, May). In for a penny: Metal theft crimes cost consumers and insurers. *Forbs Magazine.* Retrieved from http://www.forbes.com/sites/clairewilkinson/2012/05/24/in-for-a-penny-metal-theft-crimes-cost-consumers-and-insurers/

CHAPTER 6

What We Think We Know About Metal Thieves

Thus far this book has demonstrated that there are differences among scrappers (Subsistence Scrappers, Professional Scrappers, and other categories), and how scrappers are different from metal thieves (see Chaps. 3 and 4). While delineating the differences between metal thieves and scrappers is, important deeper questions remain. These questions primarily center on who metal thieves are, what their motivations for the theft are, and what techniques and methods they use to commit their crime. Each of these questions is addressed in the following chapters; however, establishing what is currently known about metal thieves is important.

Currently, there is very little known about the persons involved in metal theft. This is primarily due to the lack of data, which has led to an absence of criminological research on the topic (Bennett, 2008b; Posick, Rocque, Whiteacre, & Mazeika, 2012). This deficiency of research has caused news reports and anecdotal stories to be used as a basis for developing knowledge. As a result, theories run the gamut from drug addicts and opportunistic petty thieves to calculated, organized gangs. The following section describes what is currently known about metal thieves, several of the popularly suggested theories, and identifies the strengths and inaccuracies of this knowledge.

6.1   THE DRUG HYPERBOLE

The most commonly discussed premise about metal thieves among industry officials, law enforcement, and news reports are implications of drug usage correlating with metal theft. However, there is no conclusive or empirical

© The Author(s) 2017
B.F. Stickle, *Metal Scrappers and Thieves*,
DOI 10.1007/978-3-319-57502-5_6

evidence that metal thieves are often drug users or addicts. Despite this lack of support, news reports and anecdotal stories by industry officials and law enforcement seem to garner constant support for such a claim.

In fact, a significant portion of published reports (both governmental and non-governmental) claim metal thieves are drug "addicts who need a hit" (Berinato, 2007, p. 3). These reports tend to cite one another in a cyclical nature. Moreover, since there is a citation which purportedly lends credence to the drug claim, most readers and other authors simply assume scientific research supports the claim. The present study examined the sources and citations which most reports cite in support for the theory that drugs, specifically methamphetamine (meth), are a significant cause of the metal theft epidemic. The results identified three sources that are most frequently cited to support the claim. What is more, these reports and many others that followed cite one single and primary source. This original story appears to be the basis for nearly all claims that drug addiction is related to metal theft. This section examines each of the three most commonly cited sources on the drug and metal theft hypothesis, concluding with an examination and analysis of the original source document.

6.1.1   Government Reports

The first most frequently cited source, which claims a connection between metal thieves and drugs, is an FBI Criminal Intelligence Section 2008 bulletin on copper theft. This document is often cited as an authoritative source indicating that metal theft is conducted to support "drug addicts" (p. 2), yet the only source the FBI cited to support this claim was a single news report of copper theft in Florida. If any other data was utilized to support the conclusions in this report, they were not identified by the FBI.

The second most frequently cited secondary source, which claims a connection between thieves and drugs, is the *Problem-Oriented Guide for Police on Theft of Scrap Metal* by Kooi (2010). In this handbook, widely distributed among law enforcement agents, Kooi states,

> Drug addicts, particularly crystal methamphetamine users, appear to be linked with specific types of scrap metal theft. To support their drug habits, they require repeated and quick access to small amounts of cash, which they can easily obtain by selling small amounts of stolen scrap metal to dealers [Berinato, 2007]. However, there is little doubt that other types of drug addicts also steal scrap metal to support their habits. (p. 8)

The only support for the conclusion that meth addicts commit metal theft is a 2007 article by Berinato (discussed in the next subsection). The third most frequently cited source, which claims metal thieves are drug users, is the report presented by the U.S. Department of Energy (DOE) in 2007, titled *An Assessment of Copper Wire Thefts from Electric Utilities*. The study reports to have utilized open source news reports and "interview[s with] *a few* scrap dealers, law enforcement, and security professionals to obtain a first-hand understanding of the problem and the possible solutions" (p. 2) (*emphasis added*). Based on these interactions the study makes the following claims: "Efforts that lead to more arrests, more convictions, and stiffer penalties may reduce repeat offenders. However, these efforts will not reduce the crimes committed by methamphetamine addicts" (DOE, 2007, p. 6). This statement and claim came with no previous mention of drug usage in the report, nor were there supporting citations—media reports or otherwise—to support the allegation. Next, the DOE concludes, "There is a strong correlation between crystal methamphetamine drug abuse and reported metal theft" (p. 7).

The emphasis on meth usage continues when the report describes theft from energized substations and utility poles, stating that it is "the most dangerous place to steal copper wire," and then concludes that the theft from dangerous places is directly "related to the large numbers of methamphetamine users" (p. 8). While the DOE study does state that it "makes no claim or attempt to be comprehensive in its coverage of all copper wire thefts at electric facilities" (p. 2), the warning is not heeded as evidenced by the frequency the report is cited in support of metal theft and drug usage relationship.

6.1.2   Chief Security Officer Magazine

The primary source cited by a significant number of reports to support the claim that drug addicts steal metal to support their habit originates from a 2007 article by Scott Berinato written for the periodical magazine *Chief Security Officer* (CSO). CSO is a private organization that provides news, analysis, and research on security issues and risk management (CSO, 2007) to practitioners. Beginning in 2007, CSO took a particular interest in metal theft in Detroit, Michigan. In particular, they examined the impact of copper theft on DTE Energy Company, which provides electrical services in Michigan. In the article titled "Red Gold Rush: The Copper Theft Epidemic," Berinato interviewed half a dozen persons in law enforcement and the power utility industry in Michigan. In a subsection of the article entitled "How the drug problem got to be Mike Dunn's problem," Berinato spends several paragraphs describing the effects of meth and the activities of addicts. He explains the need for money to satisfy a drug addict's next "craving" (p. 7), the lack of sleep, high energy levels, and the tendency to complete an intensive and repetitive activity. Berinato then makes the following assertions.

First, Berinato describes a story relayed by his interviewee (Mike Dunn of American Electric Power) in which metal thieves steal copper weld grounding rods from an energized electrical substation. To obtain copper suitable for sale from copper weld grounding rods, the thieves must unravel the thin copper wire which surrounds a worthless rod. Dunn comments that it would take two hours to complete this task and adds an incredulous statement, "... for what? A hundred bucks of copper?" (p. 6). Berinato then draws the conclusion that completing this "bizarre—Herculean efforts put forth for minimal payoff ... makes sense when put in the context of crystal meth" (p. 6). However, when comparing Michigan's 2007 minimum wage of $7.15 per hour with the projected $50 per hour made unwinding copper strands from a grounding rod, it is easy to see how many persons, drug user or otherwise, might choose the latter. Clearly, this example does not support the conclusions drawn in the magazine article.

Second, Berinato utilizes several uncited news stories from around the world which describe metal theft of varying quantities and methods, including stealing 400 feet of aluminum bleachers, taking 4400 feet of wire off an electrical pole, and completely removing a 36-foot vehicle bridge, leaving an entire town stranded. While some of these purported thefts are significant and unusual, others are more common. However, Berinato concludes that addicts must have perpetrated the thefts he describes because meth addicts often stay awake and active for long periods.

Finally, Berinato describes the dangerous situations in which metal theft occurs, primarily substations with energized lines and rail yards. He then postulates that meth addicts experience "a craving so intense that they will take extreme measures ... to get more" and that "a crystal meth addict, whether high or craving a high, isn't rational about what constitutes risky behavior." He concludes this statement by describing several anecdotal stories that he admits are "wildly risky metal thefts that lead to death [and] are legion and often harrowing" (p. 7). Again, criminals taking risks to obtain financial rewards is hardly new, nor is this the exclusive domain of drug users.

To Scott Berinato's credit, he does include this caveat in the midst of his conclusions on the connection between drug usage and metal theft.

> It's important to point out that not all meth addicts are metal thieves and, likewise, not all metal thefts track back to meth addicts. No scientific data exists yet that confirms the link between the two, but CSOs and law enforcement say the link exists. Many interviewed for this story mentioned the drug unprompted. (p. 6)

Unfortunately, this statement received scant attention by those who cite his study to support the notion that drug usage and metal theft are connected.

While blame should not be placed on Scott Berinato (a journalist who reported his conclusions and observations in an industry periodical), there are significant issues with official reports and studies which cite his report as an authoritative and conclusive study on the connection between drug usage and metal theft. For example, the DOE report cites Berinato's 2007 article as the source for their conclusion, and in discussing Berinato's article in the footnotes says, "according to an extensive study sponsored by the Chief Security Officer website…" (p. 7). However, Berinato's research was not comprehensive, nor does he claim it to be. Unfortunately, Scott Berinato's magazine article has been used inappropriately for years to support the allegation that metal theft is connected to drug addicts. Berinato's article was based on anecdotal observations and a few interviews with industry officials and should not be utilized within professional research.

6.1.3   Senate Subcommittee Hearing

The hysterical hyperbole of drug usage and metal theft was advanced even further during a 2009 United States Senate Subcommittee hearing on metal theft, where all but one presenter discussed the relationship between drugs and metal theft without citing a single source for their conclusion. For example, Senator Oren Hatch claimed that metal theft is "primarily for drugs" (p. 4). Furthermore, the now defunct Coalition Against Copper Theft claimed that there was "a clear and definitive link between stealing copper and illegal drug use, primarily methamphetamines" (p. 25). Perhaps even more outrageous are statements by Mona Dohman, a police chief in Minnesota, who discussed the fallacious drug–metal theft link and also stated, "[metal theft] can be and is a gateway to further reaching and more severe crime" (p. 8). Whereby she co-opted a famous phrase of government officials to describe low levels of drug usage (e.g., marijuana) act as a "gateway" to more serious crimes, and implied not only that metal theft was the result of drug usage, but also that addicts would move from metal theft into crimes that are more serious.

6.1.4   National Insurance Crime Bureau

To date, the only known study on the relationship between metal theft and drug usage was conducted by the NICB in 2012 (Kudla, 2012). The study compared the rate of insurance claims for metal theft within each state, per 10,000 residents, with the 2009 state estimated rates of drug abuse/dependence individuals, per 10,000 residences. The NICB findings indicated that states with higher rates of substance abuse and dependency experienced higher rates of metal theft, claiming the correlation was .263 (significant at the .063 level). Unfortunately, the NICB did not provide any additional details of its statistical analysis or the raw data to support their findings. The study concluded that "the thieves are often drug addicts and steal these materials to sell them to scrap dealers and net themselves some quick cash" (p. 2). The NICB concludes the short section on drug usage and metal theft by saying, "Of course, many factors are contributing to the metal theft rates of a given area. Drug abuse may not be the primary factor influencing metal theft, but a correlation was found to exist" (p. 6).

6.1.5   Drug Hyperbole Conclusion

It is doubtful that a firm connection between drugs and metal theft has been established in the literature. However, this is not to say that there is no link, but rather to demonstrate what occurs when a lack of knowledge exists. Even a magazine article based entirely on media reports, anecdotal interviews with only a handful of persons in law enforcement, and broad correlations between state drug usage and metal theft quickly becomes the standard by which others report on the issue.

It is also important to address the repeated reference not only to drug usage but also specifically to meth. It would appear, based on the studies discussed above, that the drug–metal theft hypothesis, precisely the emphasis on meth, may be related to moral panic and hysteria, which has been driven by the media and government. The hysteria over meth use appears to have been in full swing in the United States in the mid to late 2000s (Hart, Csete, & Habibi, 2014), which coincides with the sharp rise in metal values and media attention to metal theft. Researchers have documented the effects of meth hysteria creating a moral panic (Chenault, 2012; Hart et al., 2014; Jenkins, 1994; Weidner, 2009). It is also unlikely that the estimated 500,000 methamphetamine users across the United States in 2007 (Substance Abuse and Mental Health Services Administration, 2012) could have been responsible for the dramatic rise in metal theft rates.

While a connection may indeed exist between drug usage and metal theft, one has not been established in the literature. Moreover, it is doubtful that the relationship is as solid as emphasized, and may coincide with the hysteria about meth usage. Conversely, perhaps meth users are easier to catch, and therefore are over-represented within law enforcement contacts, thereby enhancing the frenzy surrounding drug use and crime. Regardless of the reasons or source of the drug hyperbole, one thing is sure: until rich data is produced, which is scientifically analyzed and combined with evidence-based knowledge of offenders, the relationship between drug usage and metal theft within the criminological community is uncertain.

6.2 OPPORTUNISTIC THIEVES

Despite a significant number of authors, researchers, and officials who cite drug addiction as the primary motivating factor for metal theft, others believe metal theft is merely a crime of easy opportunity and profit (Bennett, 2008b; Kooi, 2010). Opportunistic crimes are those that occur with little or no planning or premeditation, as conditions are present that allow the crime to happen with little effort or risk (Clarke, 1997). This concept has been applied to many types of crime, especially property crime, and may be a factor with metal theft. Describing metal thieves as opportunistic does not necessarily preclude them from having committed a planned or organized metal theft in the past, nor does it indicate that they lack the skills necessary to complete the crime; rather, it exemplifies the mindset of the offender at the time of the theft. Kooi (2010) believes that opportunistic metal thieves steal metal when guardianship is lacking. In other words, opportunistic metal thieves may not be seeking to steal metal but may discover it during their routine activities at the places they visit and may decide to take it to obtain quick cash. Opportunistic thieves may be contractors, construction workers, or juveniles who occasionally steal and sell metal when the conditions are exceptionally easy (Kooi, 2010).

Other individuals may commit theft when the opportunity presents itself. Easily accessible metal items such as bicycles, ladders, aluminum siding, copper pipes on a parked work truck, and other unsecured metals may be a tremendous temptation for many who are presented with such opportunities. An illustration may be a subject walking through a neighborhood who observes copper pipes stored behind a shed. While this individual may not regularly be involved in metal theft (or any other type of crime), when the opportunity to quickly acquire a high-value metal with low risk and numerous accesses to recyclers presents itself, the temptation to transform metal into a quick buck may lead to theft. This type of opportunistic theft requires little knowledge or skill, and may frequently occur with small amounts of metal.

While it is certainly possible that some metal thieves lack any ability or knowledge and tend to rely exclusively on chance opportunities to locate discarded bicycles, ladders, and other forms of metal, others demonstrate tremendous capacity and expertise in the field. In fact, three abilities are needed to procure many types of metal: skills (e.g., how to identify and acquire valuable metal), the expertise to remove it, and physical means to transfer it to a buyer. Electricians, construction workers, plumbers, HVAC workers, and others both currently and formerly employed have all three of these abilities. Moreover, since these individuals have frequent contact with valuable metals, they may be presented with opportunities to steal metal in conjunction with their professions (Bennett, 2008b; ISRI, 2011). For example, a plumber may be presented with the opportunity to recycle copper pipes remaining after a job rather than return them for a refund for the owner, or a HVAC installer may observe valuable metals he can take during an installation job. Metals can be easily acquired, and because these individuals work in an industry where they are expected to have significant amounts of metal, they are viewed with less suspicion.

Since the steep rise in metal prices, many individuals have become aware of the value of the metal, the abundance of available sources, and the reduced risk involved. While most people may not routinely commit metal theft, it is likely that others will when the opportunity presents itself. Opportunistic thieves may range from unskilled individuals to those licensed in an industry related to metal (e.g., plumbers). It is important to remember that many instances of metal theft may be opportunity-driven.

However, it should not be concluded these crimes are less rational. Again, what is commonly known about metal thefts is gleaned through news reports. Consequently, unsensational cases of metal theft are less likely to make news reports, which may account for the underrepresentation of these types of opportunistic crimes in the news media and thus in present articles.

6.3   CALCULATED THIEVES

While there is evidence that metal thieves may steal based on the easy opportunity to do so, researchers and law enforcement officials increasingly point to the calculated and organized nature of many metal thieves (Committee on the Judiciary, 2009; FBI, 2008; Wonder, 2008). Calculated thieves create an important distinction from opportunistic thieves. Calculated thieves seek out any opportunities to steal metal, commonly plan the theft, often utilize specialized experience and tools, and frequently operate within a group of other criminals. Calculated thieves are difficult to study, as they often operate with efficiency, making them more problematic to identify and apprehend.

Currently, little is known of calculated metal thieves other than assumptions by researchers, industry officials, and law enforcement officials. Some of these assumptions are based on analysis of crime patterns. For example, Whiteacre, Terheide, and Biggs (2014) discovered a 25% increase in the theft of larger appliances between 2008 and 2012 in Indianapolis. They concluded that appliance theft is an indication that metal theft is becoming "more organized and purposeful than the odd thief with a shopping cart or backpack" (2014, p. 2). Other researchers (Sidebottom, Ashby, & Johnson, 2014) have noted that groups of organized thieves likely committed many instances of metal theft from the British National Railway. Posick et al. (2012, p. 94) discovered a significant relationship between abandoned properties and metal theft in Rochester, New York, concluding that metal theft is "not likely as part of an afterthought during a non-metal theft episode." Moreover, The Eau Claire Wisconsin Metal Theft Initiative (2008) also found that a significant portion of metal thieves operated in groups that appeared to be calculated and that tended to strike the same place repeatedly if sufficient metal was present. The repeated victimization of a single place also indicates that metal thefts may be calculated (Ashby, Bowers, Borrion, & Fujiyama, 2014; Whiteacre, Terheide, & Biggs, 2015).

While plumbers, electricians, HVAC technicians, contractors, and others were also discussed as possibly being opportunistic, there are some indications that a handful of these individuals frequently and deliberately steal metal. For example, a HVAC technician may scout places when legitimately working on a job and return later with the knowledge and tools necessary for a quick theft. If confronted, such thieves may even be able to present themselves as repairing a HVAC unit and be free to operate without scrutiny. These thieves are then able to circulate stolen and non-stolen material for sale at the recycling center with impunity. In fact, many laws aimed at curbing metal theft exempt plumbers, contractors, electricians, HVAC technicians, and others from many of the regulations because they are deemed to be legitimate.

Further conclusions that metal thieves are calculated are also frequently based on the volume and details of news stories. Many of these stories demonstrated the sophistication, organization, and calculated nature of metal thieves. Based on metal theft news stories, the FBI concluded in its Intelligence Assessment of Copper Thefts (2008),

> Copper thieves are typically individuals or organized groups who cooperate independently or in loose association with each other and commit thefts in conjunction with fencing activities and the sale of contraband. Organized groups of drug addicts, gang members, and metal thieves are conducting large-scale thefts from electric utilities, warehouses, foreclosed or vacant properties and oil well sites for tens of thousands of dollars in illicit proceeds per month. (para. 4)

Finally, there is limited and conflicting evidence that some portion of metal theft may be related to organized by criminal syndicates (FBI, 2008). Terri Wonder (2008) examined news stories and court documents and located several legal cases in Canada associating copper theft with Outlaw Motorcycle Gangs and Organized Criminal Syndicates based out of Russia. Wonder also identified a case in the Unites States involving an organized crime group convicted of extortion and arson conspiracies against several recycling centers after their involvement with an organized crime syndicate. Despite these findings, research by Matthew Ashby (2016) demonstrates that very few persons (0.5%) captured stealing metal from the UK railway were connected with an organized crime group. Ashby concluded that police were overestimating the involvement of organized crime groups.

Based on police reports, analysis of news stories, and claims by industry officials, there appears to be a significant portion of metal thieves who operate in a calculated manner. This purposeful behavior presents a significant challenge to law enforcement and researchers, as calculated criminals are more likely to steal large amounts of valuable metal (Kooi, 2010); often seek large sources of metal, such as construction sites, utilities, abandoned structures; and are difficult to locate. Beyond the theory that some metal thieves operate in an organized and calculated manner, the current literature does not provide adequate details on who these thieves are, their background, or how they work.

6.4   DRUG USER, OPPORTUNISTIC, OR CALCULATED?

Categorizing metal thieves into one of these categories is difficult for several reasons. First, the lack of data makes any proposed theory difficult to validate. Second, the area of metal theft and the wider area of scrapping (non-theft metal recycling) have significant variance. The individuals involved are not homogeneous and often represent various backgrounds, levels of sophistication, and modus operandi (Bennett, 2008b). Third, there appears to be significant hyperbole, especially connected with the moral panic on drug usage, which makes it difficult to delineate facts from fears. Fourth, it is likely that few metal thieves exclusively remain in a single category.

Finally, metal theft, unlike most other property crimes, is generally making it difficult to take existing categories and apply them to this area. Ashby et al. (2014) suggested that "more certain and more detailed results on the extent to which metal thieves plan their offending could be obtained using alternative methods such as offender interviews" (p. 18). This chapter and the subsequent ones provide an initial theoretical explanation of metal thieves by utilizing the proposed interview method, which provides a thorough understanding of metal thieves and a basis for continued evaluation by other researchers.

6.5   PRICE-THEFT HYPOTHESIS

When examining metal theft, it is important to understand that it is a unique category of crime (Sidebottom, Belur, Bowers, Tompson, & Johnson, 2011). Stolen metals rarely hold any intrinsic value to the thief. Something of intrinsic value has an inherent worth, which is often borne out by the pleasure a person receives from possessing the item. Examples of commonly stolen items of intrinsic value are money, cigarettes, electronics, vehicles, and clothes. Conversely, the value of metals to the thief is realized only when the metals are sold to a recycling center, which then exchanges the metal with money.

The exchange of metal for money has been referred to as "criminal alchemy" (Klobuchar, 2009, p. 2), which is a pun referring to the process by which the metal is transformed into profit. The concept of theft occurring exclusively due to the items' extrinsic value may be relatively unique to metal theft. There does not appear to be a criminological theory or research that directly addresses this concept. It is perhaps this exclusive extrinsic value that has caused metal theft to be examined through the lens of economic evaluations (Sidebottom et al., 2011), which are better equipped to analyze extrinsic values than many criminological theories.

The common assumption by many in law enforcement (Bennett, 2008a; FBI, 2008; Kooi, 2010), the metal industry (DOE, 2007, 2010; Southwire Company, 2012), and researchers (Lipscombe & Bennett, 2012) is that when the value of copper and other base metals increases the value of scrap metal is also elevated, which leads to a financially rewarding equilibrium and proves to be an impetus for the theft of metal. According to Kooi (2010), "The rise in scrap metal theft is driven by [the] offenders' recognition that ample metal supplies remain unguarded, and that the price of return remains historically high" (p. 4). This supply-demand inequality provides the opportunity to sell metals (stolen or otherwise) at a significant profit to recyclers. In other words, when supply is equal to demand, or supply is greater than demand, the price of scrap metal falls, which results in decreased metal theft. Whereas when demand is high and supply low, metals increase in value and thus are a more attractive target. This concept is anecdotally supported through observation by Roggio (1998), who discusses a decrease in participation with the California Metal Investigators Association because "the group is finding now that some metal prices are so depressed that incidents of scrap theft are down considerably" (p. 5). This decline of interest in the organization occurred in 1998 as the price of copper plummeted from an all-time high, in 1995, of $1.38 per pound to a decade low of $0.78 per pound.

6.5.1   National Insurance Crime Bureau

Perhaps the first organization to publish findings on the price-theft hypothesis during the current metal theft trend was the NICB. The NICB discovered a 3000% increase during the first six months of 2008 in insurance claims due to stolen catalytic converters. NICB also identified a 55% increase in the price of rhodium and a 27% increase in the price of platinum, both materials commonly found in catalytic converters. Unfortunately, the hypothesis was never statistically validated, and NICB did not include sufficient data in their report to conduct a third-party analysis. Moreover, no other organization, including the NICB, has continued to examine this trend. Presently, the only available data analyzing catalytic converter thefts is the 2008 NICB study.

The following year the NICB identified a "consistent" (Kudla, 2012, p. 6) correlation between non-automotive insurance claims of copper theft and copper prices in a study they released in 2009. This study examined insurance claims for metal theft between January 2006 and November 2008. Once again, this report did not indicate the strength of the correlation. However, a third examination, published in the beginning of 2012, did provide a statistical analysis of the price-theft hypothesis. Examining metal theft claims from January 2009 through December 2011, the report found "a statistically significant correlation with the price of copper" (Kudla, 2012, p. 1). While the report did not provide the full analysis, it did indicate that the "Pearson's correlation coefficient … was .903 [which is] significant at the .001 level" (p. 1). A follow-up study conducted from 2011 to 2013 also claimed to support the price-theft hypothesis; however, no statistical data or support was provided (NICB, 2014).

6.5.2   Rochester, New York

Research has also demonstrated support for the price-theft hypothesis in local markets. Chad Posick, in conjunction with the Center for Public Safety Initiatives, examined copper-related burglaries in Rochester, New York, between January 2007 and November 2008. The findings indicated "a moderately strong relationship" (p. 2) between copper prices and theft (Posick, 2008). This study also examined non-copper-related burglary trends and concluded that the price of copper was a better predictor of metal theft than the general burglary rate. The study also indicated that there was a two-month lag in market copper prices and the rate of copper burglaries. These findings demonstrate the strength of the price-theft hypothesis, as it would likely take time for the global price of copper to affect the local market and then for thieves to respond.

In 2012, Posick et al. expanded the study of copper-related burglary in Rochester even further. That study examined burglary data from April 2008 through July 2010. Of the 5656 burglaries investigated, 585 were identified as metal (copper) theft related. Among other analyses, the authors conducted a bivariate correlation analysis of the price-theft hypothesis and found a correlation between metal prices and copper burglaries of 0.73 ($p < .01$), which "indicated a close relationship between the two" (p. 93). This relationship was stronger than any other data examined (e.g., trends in non-copper-related burglary, seasonality), suggesting that the price of metal is the most significant predictor of theft.

6.5.3   British Railway Network

The largest and the most exhaustive study to examine the price-theft hypothesis was conducted with data from the British Railway Network. This network includes more than 21,000 miles of railways and 3000 railway stations in the United Kingdom. Sidebottom et al. (2011) examined 2870 recorded instances of metal theft from January 2004 to October 2007 and conducted a regression analysis with copper prices reflected from the London Metal Exchange, the national crime rate, and national unemployment rates. The initial results indicated a 649% increase in copper thefts from 2005 to 2006, which coincided with the dramatic rise in copper prices discussed above. The results from the ordinary least squares (OLS) regression demonstrated a positively correlated relationship between theft and the price of copper, with a significance of 0.0001. The other two comparisons, crime rates and unemployment rates, had little to no significant relationship. Sidebottom et al. (2011) concluded: "Though causation cannot be inferred … the findings are consistent with the proposal that copper has become an attractive target for theft due to the higher prices of copper, and that such price shifts may have increased the opportunities for offenders to achieve financially rewarding sale prices" (Sidebottom et al., 2011, p. 408).

In 2014, Sidebottom et al. tested the price-theft hypothesis again with a longer series of data, from January 2006 to April 2012. The results from the study again supported the price-theft hypothesis. The study indicated movements in the average monthly price were significantly associated (<.05) with monthly energized copper cable theft. This correlation remained "almost identical" (p. 694) to changes in the value of copper, which fluctuated significantly from 2006 to 2012—between $4.59 and $2 per pound. In other words, this research demonstrated that even a monthly drop in the price of copper resulted in a decrease in theft. Further, two alternative hypotheses—unemployment and police investigation—were found to have no relationship with the rate of thefts.

6.5.4   Price-Theft Hypothesis Conclusion

Much of the empirically tested evaluations of any aspect of metal theft appearing in the literature are research based on the price-theft hypothesis. Moreover, only three of the assessments, Sidebottom et al. (2011), Posick et al. (2012), and Sidebottom et al. (2014), have been accepted in peer-reviewed journals. The rest of these evaluations is published reports from various governmental and non-governmental organizations. However, the research and findings of the price-theft hypothesis indicate several important issues. First, metal theft is tough to study, primarily due to a lack of data. The rates of local metal theft (Rochester and Indianapolis), or through one company (The British Railway Network), or through insurance claims (NICB) represent some of the only data available on metal theft. Therefore, without rich data on the thefts, victims, and criminals, much of the research on metal theft will continue to be an evaluation of the price-theft hypothesis. While it is important to keep this issue under review, it demonstrates the need for considerably more data on metal theft so that additional avenues of examination can be developed.

Second, the price-theft hypothesis between the rate of metal theft and the price of metal (copper in particular) has been statistically validated by several studies and across national, local, and international areas. This correlation remains strong despite significant monthly fluctuations in the value of the metal. The relationship marks a distinct departure from many other types of crime and is likely related to the extrinsic value of metal (Sidebottom et al., 2011).

Finally, the validation of the price-theft hypothesis has significant impacts on the current and future evaluation of crime prevention efforts, specifically when examining metal theft prevention efforts, including law enforcement efforts, regulations, or legislation. If the price of copper is not included in a statistical analysis of these efforts, the results indicating that the efforts or laws are successful may be spurious. Future research needs to evaluate crime prevention efforts, while controlling for the price-theft hypothesis to ensure that accurate measures of prevention techniques occur.

6.6   SPATIAL ANALYSIS

In addition to the scholarly research conducted around the price-theft hypothesis, there is a growing body of empirical research identifying how the built environment or place impacts metal theft. However, rather an address this research here, it is included in Chap. 8, which is largely devoted to how place influences metal thieves.

6.7   SUMMARY

This chapter has drawn on all the available research, both published and unpublished, scholarly and non-scholarly, to examine what is currently known about metal thieves. Unfortunately, firm conclusions are difficult to establish, given the varying nature of the data analyzed, the data source, and the methods used to examine them. Much of what is "common knowledge" about drug usage and the connection between metal thefts appears to be anecdotal at best and fallacious at other times. Currently, there is no strong connection between metal theft and drug usage; rather, the same reports citing the connection seem to be recycled time and again, thereby creating the drug hyperbole.

The limited research on the activity of metal thieves is split on thieves being calculated or opportunistic. However, it is unlikely that these categories are mutually exclusive; rather, metal thieves can be both organized at times and opportunist at others. Hence, even a criminal who frequently organizes and calculates a metal theft may still steal metal if the opportunity allows. There is some evidence to support the notion that metal theft often occurs in a team or group of criminals, yet it is doubtful that these groups are organized crime groups or syndicates.

The strongest research currently published on metal theft indicates that there is a significant relationship between the price of copper (used as an index of general metal value) and the rate of metal theft. This connection has been statistically demonstrated in two United States cities and several times in the United Kingdom. In fact, the price of metal has been the strongest predictor of metal theft and is the only theory validated by peer-reviewed empirical research.

REFERENCES

Ashby, M. P. (2016). Is metal theft committed by organized crime groups, and why does it matter? *Criminology & Criminal Justice, 16*(2), 141–157.

Ashby, M. P., Bowers, K. J., Borrion, H., & Fujiuama, T. (2014). The when and where of an emerging crime type: The example of metal theft from the railway network of Great Britain. *Security Journal*, 1–23.

Bennett, L. (2008a). Assets under attack: Metal theft, the built environment and the dark side of the global recycling market. *Environmental Law and Management, 20*, 176–183.

Bennett, L. (2008b). *Metal theft: Anatomy of a resource crime.* Unpublished. Retrieved from http://shura.shu.ac.uk/4125/

Berinato, S. (2007, February). Red gold rush: The copper theft epidemic. *Chief Security Officer.* Retrieved from http://www.csoonline.com/article/2121953/loss-prevention/red-gold-rush

Chenault, S. (2012). New ice age: A Content analysis of methamphetamine coverage in a Midwestern newspaper. *Journal of the Institute of Justice and International Studies, 12*, 15.

Chief Security Officer. (2007, February). *Chief Security Officer.* Retrieved from http://www.csoonline.com/about/about.html

Clarke, R. V. (1997). *Situational crime prevention: Successful case studies.* Guilderland, NY: Harrow and Heston.

Committee on the Judiciary. (2009). *Metal theft: Public hazard, law enforcement challenge. Hearing before the Subcommittee on Crime and Drugs of the Committee on the Judiciary United States Senate One Hundred Eleventh Congress.* Washington, DC: U.S. Government Printing Office.

Eau Claire Police Department. (2008). *The Eau Claire, Wisconsin metal theft initiative: Crime facilitators—Where the problem meets the solution.* Eau Claire, WI: Eau Claire Police Department, Detective Division.

FBI. (2008). *Copper thefts threaten U.S. critical infrastructure.* Washington, DC: FBI Criminal Intelligence Section.

Hart, C. L., Csete, J., & Habibi, D. (2014). *Methamphetamine: Fact vs. fiction and lessons from the crack hysteria.* New York: Open Society Foundations.

Institute of Scrap Recycling Industries, Inc. (2011). *Preventing metal theft.* Washington, DC: Institute of Scrap Recycling Industries, Inc.

Jenkins, P. (1994). "The ice age" the social construction of a drug panic. *Justice Quarterly, 11*(1), 7–31.

Klobuchar, A. (2009, July 22). Interview by L. E. Long. Metal Theft: Public Hazard Hearing.

Kooi, B. (2010). *Theft of scrap metal: Problem-oriented guides for police series. Guide No. 58.* US Department of Justice, Office of Community Oriented Policing Services.

Kudla, J. (2012). *Data analytics forecast report: Metal theft claims and questionable claims from January 1, 2009 to December 31, 2011*. Des Plaines, IL: National Insurance Crime Bureau.

Lipscombe, S., & Bennett, O. (2012). *Metal theft*. London: House of Commons Library.

National Insurance Crime Bureau. (2014). *NICB: Insured metal theft claims see three-year decline*. Des Plaines, IL: National Insurance Crime Bureau.

Posick, C. (2008). Examining the relationship between copper burglaries and the price of copper in Rochester, NY: Center for Public Safety Initiatives: Rochester Institute of Technology.

Posick, C., Rocque, M., Whiteacre, K., & Mazeika, D. (2012). Examining metal theft in context: An opportunity theory approach. *Justice Research and Policy, 14*(2), 79–102.

Roggio, A. (1998, October 14). Thieves find some scrap not worth stealing. *American Metal Market*, Vol. 106.

Sidebottom, A., Ashby, M., & Johnson, S. D. (2014). Copper cable theft revisiting the price–theft hypothesis. *Journal of Research in Crime and Delinquency, 51*(5), 684–700.

Sidebottom, A., Belur, J., Bowers, K., Tompson, L., & Johnson, S. D. (2011). Theft in price-volatile markets: On the relationship between copper price and copper theft. *Journal of Research in Crime and Delinquency, 48*(3), 396–418.

Southwire Company. (2012). *Curbing the copper theft crime wave*. Arlington, VA: Rural Electric Magazine. Retrieved from http://remagazine.coop.

Substance Abuse and Mental Health Services Administration. Results from the 2012 National Survey on Drug Use and Health: Summary of National Findings, NSDUH Series H-46, HHS Publication No. (SMA) 13-4795. Rockville, MD: Substance Abuse and Mental Health Services Administration, 2013.

U.S. Department of Energy. (2007). *An assessment of copper wire thefts from electric utilities*. Washington, DC: Infrastructure Security and Energy Restoration: Office of Electricity Delivery and Energy Reliability.

U.S. Department of Energy. (2010). *An updated assessment of copper wire thefts from electric utilities*. Washington, DC: Infrastructure Security and Energy Restoration: Office of Electricity Delivery and Energy Reliability.

Weidner, R. R. (2009). Methamphetamine in three small Midwestern cities: Evidence of a moral panic. *Journal of Psychoactive Drugs, 41*(3), 227–239.

Whiteacre, K., Terheide, D., & Biggs, B. (2014). *Research brief: Metal thefts in Indianapolis October 1, 2011–September 30, 2013*. Indianapolis: University of Indianapolis Community Research Center.

Whiteacre, K., Terheide, D., & Biggs, B. (2015). Metal theft and repeat victimization. *Crime Prevention & Community Safety, 17*(3), 139–155.

Wonder, T. K. (2008). Organized crime and metals theft: A "premonitory" model for investigators and analysts. *IALEIA Journal, 18*(1), 69–82.

Metal Thieves in Their Own Words

PART III

Metal Thieves in Their Own Words

CHAPTER 7

Metal Thieves

As has been previously demonstrated in the findings related to taxonomy and the scrapping subculture, metal thieves are unlike scrappers in many ways; metal thieves do not share the same values, codes, norms, and beliefs as scrappers. Consequently, it is important to distinguish who metal thieves are. This chapter delves into what is known about metal thieves, seeking to establish metal thieves' common characteristics, actions, beliefs, and motivations.

7.1 DEFINING METAL THEFT

Much of the research on metal theft tends to assume an unwritten definition or relies on a local or state ordinance to define metal theft. There are three significant problems with this method. First, local ordinances and state laws may vary significantly, thus making metal theft difficult to define, track, and study across varying geological boundaries. Second, some definitions fail to encompass the scope needed to understand all the components of metal theft. For example, some definitions and legislation on metal theft often use the term scrap metal. However, scrap metal typically means a metal that has come to the end of its useful working life. Therefore, and perhaps inadvertently, some definitions and laws related to metal theft only include metal that is no longer useful. These definitions fail to consider metal theft of new copper pipes, vehicles in drivable condition, copper wires currently in use, air conditions, or any other metal item that is still performing its intended function.

© The Author(s) 2017
B.F. Stickle, *Metal Scrappers and Thieves*,
DOI 10.1007/978-3-319-57502-5_7

The third problem with many definitions of metal theft is exactly the opposite of the first two. In this case, the definition may be too ambiguous or broad in scope. For example, some cities have ordinances that stipulate that any waste set out at the curb for recycling or solid waste collection becomes the property of the city. In other words, the city becomes the legal owner of trash or waste when it is moved to the curb by the owner. Therefore, any unauthorized removal of the waste, metal or otherwise, is a criminal offense. Strict application of this definition would result in defining nearly all Subsistence Scrappers, Professional Scrappers—or even homeowners who change their mind about an item they set out as trash— as criminals and metal thieves. Clearly, defining metal theft this broadly unnecessarily widens the scope of metal theft and alters the focus of prevention efforts.

Due to these complexities, developing a definition that encompasses all the necessary components of metal theft without being too inclusive is difficult. For the present study, a definition of metal theft used by Whiteacre et al. (2014) is employed:

Metal Theft: "the theft of item(s) for the value of the constituent metals."

However, an important caveat will be the exclusion of those who may technically be committing theft (due to local laws) when they acquire metal items that have been set at the curb for disposal. These articles set out as trash have historically been considered free to take by both the previous owner and the individual collecting the items (Ferrell, 2006; Zimring, 2009), and this is not the focus of the present study. This exclusion will enhance the ability to concentrate on metal thieves who acquire metal from individuals who maintain affirmative possession of their metals.

7.2   WHO ARE THE METAL THIEVES?

The sample size and geographic location of metal thieves in the present study are not robust enough to provide generalizable demographics for all metal thieves. Moreover, several of the thieves interviewed declined to provide personal information (e.g., age, education), citing privacy concerns. However, based on the demographic data available, the metal thieves in the present study tended to be male (92%) and younger than scrappers, with the most common age range of 30 years followed equally by those in their 20s and 40s. Moreover, while blacks and Hispanics are represented, the majority of the metal thieves were white (85%). Interestingly, of those who agreed to answer questions about education, 54% of metal thieves had some level of a college experience. The high level of education among metal thieves exceeds the 10% of scrappers who stated that they had college experience; however, this is consistent with the national average of 59% of persons aged 25 years and older who have some college experience (US Census Bureau, 2015).

The over-representation of college experience among metal thieves compared to scrappers may be a result of employment and advanced technical training among many metal thieves. The majority of metal thieves (69%) were employed full-time while stealing. Moreover, all of those employed had past or current work experience in an industry that granted access to, training in, and valid justifications for possessing metals. For example, metal thieves' professional experience ranged from employment at a recycling center, electricians, HVAC installers, work in general maintenance, construction contracting, and even as a car salesman who specialized in taking junk cars on a trade. Due to the high frequency of employment, along with college experience, it is not surprising that less than 15% of metal thieves in the present study were homeless. Moreover, drug usage was a factor discussed in only about 30% of the metal thieves' interviews.

These findings stand in stark contrast to the prevailing opinion among law enforcement, which is typified by a meeting with Detective Baker, a field detective who frequently investigates metal theft on the West Coast:

Detective Baker: Most of the people that I deal with in reference to metal theft, they are methamphetamine or heroin users, [with] very little to no education at all. That is something that intrigued me when I got into the metal theft investigation aspect of it, [that] the highest level of education of the people that are arrested is 7th or 8th grade. Most of the people that violate the metal theft laws are male, but you have a high number of females who are stealing it now for the simple fact that it's easy money… Usually, for the most part, the typical scrapper the typical metal theft thieves are going to be homeless people between the ages of 35 and 55.

Interviewer: Okay, well, these are pretty specific numbers. Is this something you thought about before and looked at it?

Detective Baker: Yeah, I am actually interested in developing maybe a company largely to study metal theft and how to help prevent it, so I've done a lot of research (reading periodicals and new reports). Because of all of the metal thieves that [I know] those are the [factors] that I've found to be the most common.

It is challenging to identify the reasons why the experiences of Detective Baker and many others in law enforcement contradict the findings in the present study. The differences may be due to location; while seven states are represented in the present study, only one metal thief was interviewed from the West Coast, where Detective Baker worked. More likely, the discrepancy may exist between individuals who are often captured by law enforcement while stealing. Only about half (53%) of the metal thieves in the present study had been arrested for metal theft. Therefore, the difference between common perceptions by law enforcement and the findings in the present study may indicate that metal thieves with college experience, who do not use drugs, and who are employed are less likely to be caught than other criminals who lack education, are homeless, and are drug users.

7.3   MOTIVATION FOR THEFT

After establishing general demographics of metal thieves within the present study, it is important to evaluate motives. Understanding how and why individuals began to commit theft, why they continue stealing, and primarily why metal is targeted over other items is an important and rarely examined issue. Gaining a firm appreciation of the motivations of metal thieves will assist in understanding why and how they operate and what techniques can be utilized to prevent metal theft.

7.3.1   Scrappers Drifting into Theft

Kevin Whiteacre, Associate Professor at the University of Indianapolis and Director of the Indianapolis Metal Theft Project, hypothesized that the majority of thieves begin honestly scrapping, but are drawn into a gray area and start to break the law (Whiteacre, 2014). Whiteacre's comments are supported by the present study, which found that nearly all of the metal thieves began scrapping legally before engaging in theft. Very few started recycling by stealing; rather, they tended to drift into metal theft. A criminal drifting into crime is not a new concept. David Matza (1964)

developed the theory that criminals drift out of normal behavior, tempo-
rarily setting aside conventional moral norms, values, and codes. Similarly,
the drift into metal theft tended to occur in one of two ways.

The first and most common drift into metal theft was when a friend
or acquaintance took a scrapper along to steal metal. During this first
venture into metal theft, the once lawful scrapper realizes the ease and
significant amount of money involved in metal theft. Eric described the
first few weeks he was involved with metal theft after a lifelong friend, who
had been stealing metal for years, sought him out as a partner, "You know,
honestly, you almost got hooked on it like a doggone drug or something
because … honestly, to me it was like thrilling, easy money!" Similarly, Leo
summarized how he began his foray into metal theft:

> I was doing it legally, and then a friend of mine told me about, you know,
> abandoned buildings, you know, copper and stuff like that, and it sure was
> worth a lot of money! I didn't believe him, but I did it the first time and
> made a substantial amount of money, and I was like, what the…, and it kind
> of went from there. The first time I think it was like three grand! $3000!

Comparably, Matt, who is a licensed apprentice journeyman and installs
air conditioners for a local company, had been a professional who scrapped
in the course of business for years before his drift into metal theft. He
related his first experience with air conditioner thefts, describing how his
partner at the HVAC company they worked at encouraged him to steal air
conditioner units after work.

The second most common drift into theft involves an individual who
begins slowly crossing the boundary from legal into illegal while scrapping.
David explained how easy it is to cross the line from legal scrapping into
theft, commenting, "It's the temptation of ease; it's how it presents itself.
People set themselves up to have it done, because for some reason they
don't guard themselves." Other metal thieves also discussed the temptation
to steal items while scrapping, such as Zach, who said, "I see someone with
a nice metal lawn set, nobody is around, and I can run out there and grab
that, and it's gone! You know, the temptation is there." Similarly, James,
who was a Professional Scrapper, explained his entry into theft this way:

> Yeah, I stole scrap before. I know where they was setting out some rims one
> day, they was going in the dumpster anyway, they didn't want people to get
> in there, there was a no trespassing [sign]. But at nighttime, you can come
> through, well, what the hell, you know?

Likewise, John, who was a professional who scrapped while working on a farm, became frustrated with the limited availability of metal and the significant work involved in procuring it, and drifted into crime when he and his partner observed an easy target: tobacco frames (made of steel) on an adjacent farm.

> Basically, because we enjoyed the money we was getting from the scrap from stuff out of the woods and everything [from] the family's farm and ... we ran out of scrap. And, we just looked at it (the tobacco frames), oh, that would be easy to get, that would be good money! $87 (the amount each stolen tobacco frame was sold for) is hard to come by when you're getting little nitwit stuff out of the woods. You have to work 4 or 5 hours to get a trailer load ... and you only have $60 or $70 worth. So, we just liked the spare money, and we didn't really think it through ... we just kind of pulled in there and just took them.

Nearly all of the metal thieves in the present study were scrapping before stealing metal and most described situations where they drifted into criminality. On occasion, this happened gradually over time, as temptations and frustration with current metal supply or personal financial struggles grew. At some point, the struggles mounted until the once lawful scrapper succumbed to the ease and temptation by drifting into metal theft. At other times, the drift occurred more quickly and usually involved a friend, family member, or acquaintance who took the individual with them while stealing metal. Scrappers who accompanied a metal thief on a journey to take metal often expressed doubt and concern at the initial idea of theft, but drifted into metal theft activities themselves when they experienced the ease and profits of theft.

7.3.2   Criminal Enjoyment

Nearly one-quarter of all metal thieves spoke of metal theft in positive terms, describing how they enjoyed stealing or were addicted to it. When I asked how taking metal made metal thieves feel, Dustin replied, "I liked it," and Leo said, "Technically [it's] an adrenaline rush, you know? I got addicted to it because it was exciting and it was a rush, but scary at the same time." Similarly, Chris discussed his enjoyment of metal theft as he proudly led me through the interior of the abandoned building where he lives as a vagrant and which he strips of metal:

There is a whole upstairs! (Chris, excitedly, took me up the stairs and gazed, longingly, at the attic area, some 15 feet above the floor). There is wire right there. I just got to figure out how to get to it (*sotto voce*). That's the fun part! If I could get to it, I would have it. It's just climbing up there to get it.

Other metal thieves also commented that stealing metal is addictive and shared stories of how they receive a rush from their criminal activity. Jessica described what it felt like when she committed burglary of a residence to steal metal, saying,

> Well, you would get a rush … especially if the wiring hadn't already been took out. You know, especially, because the older the house, the more copper you are going to find in it. Like … there are a lot of trailers, old trailers, and a lot of the water lines are still copper. So, you can get under there and take out all the water lines. It would be a rush to see what you can find!

Dustin, who scrapped partially to support his drug habit and has been arrested for metal theft several times, compared his drug induced rush and addiction to the excitement and risks that metal theft provided:

> *Dustin*: I was addicted to it, and it did serve a purpose as well, best of both worlds!
> *Interviewer*: Would getting a big find give you a rush?
> *Dustin*: Yes, every time I did it I got a rush, the bigger the score, the bigger the rush!
> *Interviewer*: Did you enjoy it?
> *Dustin*: Yes.
> *Interviewer*: Leaving your drug addiction aside, could you have stopped scrapping if you wanted to?
> *Dustin*: I don't think I could've left it alone. I'm an adrenaline junkie, and I love the rush no matter how I get it. I do extreme things all the time.
> *Interviewer*: Did you always need the money, or did you do it 'cause you enjoyed it?
> *Dustin*: No, I didn't always need the money. I did it out of habit and [for] the rush sometimes.

Jessica, too, explained that she doubted she could stop scrapping, even if she had enough money:

> *Jessica*: Well, yeah!'Cause it's addictive! (Her eyes widening with excitement when she realized I understood the addictive nature of metal theft)
> *Interviewer*: What's addictive about it?

Jessica: Just to see what you can find and how much you can get out of it.
Interviewer: Regardless of if you're high or not, it's just fun to do?
Jessica: Yeah!

While Jessica and Dustin stole metal partially due to a drug addiction, other metal thieves explained how the crime itself was addictive. For example, Michael and Eric, who were partners, described their experiences in metal theft this way:

Interviewer: What was it like the first time?
Michael: From that day forth, my goal was to scrap anything we could find. I thought I was on top of the world from scrapping.
Interviewer: Did it give you a rush, were you addicted to it, and was it scary to do it?
Michael: It was scary taking it in for sure. It would have been a rush and a bigger deal if we [were] stealing it like a robber or something, but the way it worked at [the energy company, it] was just a common occurrence. We didn't try to hide it or anything.
Interviewer: Could you have stopped if you had wanted to?
Michael: At the time I couldn't stop. It was just too easy. A few times when I was on vacation or away from work, I remember guys texting me and telling me how much money I was missing out on, and I would freak and couldn't wait to get back to get more.

In the same way, Eric also described the excitement, pleasure, and addiction he experienced from theft, indicating it was so strong that he could not stop even when the group he stole with believed they were under investigation. When I asked Eric, "Could you have stopped if you wanted to?" Eric replied as if he had never even considered the question:

I don't know if I could have stopped, quite honestly, I can't say for sure that I could've stopped. (long pause) I mean I could not say that, I mean the only reason we stopped was because the heat was on, we slowed down. Even when the heat was on, we still tried to get a little here and there, so I guess that's certain (with a disappointed tone), no we weren't going to stop.

Leo also discussed how "It was gratifying ... we did this in two days, and I just got like $2400 bucks ... for two days of work, that's great! Yeah, you get excited. You definitely are kind of happy about it."

Despite some metal thieves citing their enjoyment and gratification from stealing and describing how it provided a high that they sought, others discussed how they did not enjoy stealing or expressed conflicting emotions about their activities. For example, when I asked Matt if he enjoyed stealing air conditioners and burglarizing homes and businesses, he said, "Sometimes, sometimes it was fun. Most of the time, no, it was just to get that money to party, to get that drug." Others, like John, described the adrenaline rush they received, but explained it in regretful tones:

> *Interviewer:* [You mentioned an] adrenaline rush. Did you kind of like that; was it fun to do it?
> *John:* Not necessarily, I think it was just ignorance, you know, like when you do something, you didn't think [it through], and then you say you should have thought about it? You do something, and it was kind of a spur of the moment type of deal ... we seen how easy the money was and just kind of—(lets out a long sigh) and figured we was the only one that knew about it.
> *Interviewer:* You were just kind of doing it for money, not the excitement?
> *John:* Yeah, I love money!

Other thieves were more ambivalent in their responses. For example, when David was asked if he enjoyed stealing, he replied, "I didn't really, no, I didn't enjoy it. I didn't dislike it." When James was asked the same question, he said, "Actually, no, I don't enjoy it, but I know a dollar has got to be made today or bills won't get paid when they need to be paid." In the same way, even Leo expressed regret for his thefts, saying, "It's not something that I wanted to do. It was something that needed to be done to survive, to pay the bills. How did I feel about doing it? Not too good, but better when I had the money."

Finally, Chris expressed exasperation and disappointment at his current state of living and the crimes he has committed to stay alive and feed his drug habit. After completing the tour of the abandoned building he was stripping and living in, we had the following conversation:

> *Interviewer:* And so your plan when this one [abandoned building] dries up is to move on to another one?
> *Chris:* (In a disillusioned, but very nonchalant tone) Yeah, probably. (After a long pause, a deep drag on a cigarette, and glancing at the five-gallon bucket he uses as a restroom) That or get my shit straight. You know what I am saying? That way, I am not living like this (slowly rotating around the room with both arms out and then letting them drop to his side in an expression of disillusionment). I am sick of it! It's hard (spoken in a dark voice).

Interviewer: So if you woke up tomorrow and had a whole bunch of money, would you still do scrapping?
Chris: No! (Responding nearly before the interviewer finished the question). No, sure wouldn't!

Metal Thief Profile

Name:	Chris
Taxonomy:	Metal Thief
Location:	Southeast—Inner City
Age:	34
Race:	White
Education:	11th Grade
Employment:	Government Assistance
Experience:	Early one morning I observed Chris walk into a scrap yard. Based on my observation several months into the project, I was immediately suspicious of the metal he presented. There was a small amount of wires, the panel from an electrical breaker box, a burglary alarm speaker, part of an overhead light, a few short stubs of copper piping, a door knob, and other strange items which netted him about $10. Chris looked a little hungover with bloodshot eyes, slightly slurred speech, and was a tad unsteady on his feet. He also had a large cut down the front of this forehead that looked as though it had been sutured within the last few hours. He explained that he woke up "strapped to the bed (in the hospital). I was like, what? Why am I strapped to the bed ... I was like, wow, what the fuck happened?" As we continued to talk, I shared a cigarette with Chris and he began to open up. He explained that he was hooked on heroin and Spice (synthetic marijuana) and began telling me the story of the downfall of his life, including his theft of metal.
	Chris would live in an abandoned building or house in the inner city and take small amounts of

metal to sell each day, earning between $5 and $20, with his best day of stealing earning $47, which would help provide him with enough to eat and some drugs. He explained that he had cleaned out three buildings so far and would continue moving on to the next when the current one dried up. On one occasion, we walked together and talked, looking at houses, and Chris explained what would make each of them a good or bad target. As we walked down the streets littered with drug paraphernalia and trash, Chris spotted a backpack lying on the porch of a house. He said excitedly, "I need a bag too. I want to snatch that." Looking around to see who was watching, he cautiously walked up to the door, looked inside, quickly grabbed the bag, and walked back toward me. He inspected the bag and then quickly folded it up so he could carry it inside his shirt, as a police car sped by with lights and sirens blaring, to which he said, "Fucking pigs." A few blocks later, we arrived at the building he was burglarizing. He stopped and faced me, leaning in a little and speaking in a soft yet excited voice with his eye wide, "You wanna go in?"

Everything in me that was still cop related screamed "No," but the curious side of me said, "Go for it!" I shook my head and he explained how we had to wait until no one was looking to sneak in the back entrance underneath a board that was loose. "Watch your step," Chris said as I entered the building and allowed my eyes to adjust to the darkness. It was very difficult to see, since all the windows were boarded up. The smell was stale and old. "This is a horrible place," Chris said, as I nodded in agreement. Chris remarked, "It's not nothing fancy, dude. I don't even sleep back here. A lot of people come in right here and shoot dope or whatever," pointing to the corner where there was an old arm chair with a small table next to it with needles, burnt spoons, and corner baggies littered around. We continued to walk down a hall, toward the front of the building. The hall was so dark I could see nothing. The floor was weak and bowed significantly under each footstep, and Chris hollered back saying, "Be sure you walk straight." It was at this point that I was again nervous. I thought, "Here I am, no one knows where I am, trespassing in a building with who knows who else inside, with a guy coming off a bender, and I can't see my hand in front of my face. Who knows what might happen to me?"

We eventually emerged into a brighter, large room and I watched Chris rip more metal from the building as he discussed his plans to steal more. Despite all of this, I had great pity for Chris. He was a smart guy, and likely capable of a lot, but he had a rough life. He was hooked on drugs and could not get his life together and was sleeping in a corner of an abandoned building in the cold winter, stealing just enough to eat and get high. Eventually the conversation turned serious when I asked if he liked living this way and stealing to survive. "No", he said very softly, shaking his head slowly as his eyes turned from excitement at the metal to a soft regret and disappointment. "I don't like living this way," he replied in a very sad and genuine tone. "But I got to survive. I got to do something to survive. You know, that is my thing. Survive, survive. Wake up do one thing that day, just focus on that day, just one day at a time. That is how I do it. Sometimes I live my life a minute at a time. You know what I am saying?"

I left, heading east as Chris went west in search of drugs. As I walked back along the streets and alleys, I had an overwhelming sense of sympathy for Chris. I had felt that before as a cop, but this time it was different. Despite his hard exterior and his burglary and theft, he hated his life, but could not get it straight. I thought back on all the folks I had come across with similar stories and felt a renewed desire to use what he had told me to prevent crimes rather than focus on punishing those who commit them.

As Cesare Beccaria said in 1764, *"It is better to prevent crimes than to punish them."*

Image 7.1   The metal items Chris brought which aroused my suspicions

Image 7.2   The abandoned building where Chris was stealing metal

A small number of thieves expressed outright disdain for their actions. The majority of metal thieves, however, fell somewhere in the middle, often vacillating back and forth between enjoyment and displeasure. They often expressed delight at the challenge, excitement from the high of a big find or the pleasure of being paid, yet they also expressed dissatisfaction with what they were doing. However, a few thieves enjoyed the excitement and rush that metal theft provided and did not appear to think often or care about the consequences of their actions. To some degree, the rewards, both monetary and emotional, may be enough to encourage metal thieves to continue despite the risks of capture and feelings of guilt, such as with Eric and Michael. When the addictive enjoyment combined with the neutralization discussed in Chap. 10, it may be difficult to dissuade metal thieves.

7.3.3   Metal Theft Profits

While to varying degrees metal thieves enjoy stealing, nearly all identified money as the primary reason they stole. In fact, a few even described how they had come to view metal as money. "I see money, I see money in that dumpster still, and here's some!" James said as he excitedly reached in and pulled out a severely damaged toy doll stroller with a metal frame. While money may be a common denominator for metal theft, it is important to understand how the money is used. The three most common uses are paying for bills or other living expenses; for drugs, alcohol, and tobacco; and as fun money.

General Expenses and Bills
The most common response from metal thieves as to what they did with the profits from their thefts was to spend the money to eat, pay bills, or on some other type of regular expenditure seen as necessary to survive. Several, like David, cited a lack of work and thus the necessity to steal because they had no income to buy necessities:

> Guys [are] doing this to feed their families and maybe they've been out of work forever … You know, but I still got to, it doesn't buy toilet paper or toothpaste and stuff like that, you know when you ain't got no money.

In the same way, Chris also mentioned being laid off and explained how he had to steal to survive and buy food, saying, "Yep, I got laid off a week ago … It's [stealing metal] not something that I do all the time, but here lately it's been rough."

For others, stealing metal appeared to be the only option to survive. James explained, "I have to … this is the only way I have to make money. I do it because I know I've got bills to pay." Similarly, while in line at a recycling center, I met Robert, who related the reason he began stealing metal when he was 17 years old:

> *Robert*: Everyone in the house (both sets of grandparents, mother, and father) either died or committed suicide in two years' time and I was stuck in the house at age 17 by myself, no parent's, no nothing, nothing! I was trying to live; it was rough, man (he says this with wide eyes, shaking his head back and forth). I was a kid after they, you know, passed away. I did anything I could to come up with money then, you know?
> *Interviewer*: So your family all passed away, and you're out scrapping to survive?
> *Robert*: Pretty much!

Robert continued his story through to the present day. He is now in his mid-20s, a successful mechanic making $22 an hour, and continues to recycle.

Interviewer: So, why are you recycling today?
Robert: I don't get paid until Monday, so if I get the shit I might as well go get an extra couple of dollars, but I don't do this for a living ... when I get a little something, something, I'll bring it down here just to make a little extra few dollars. I don't do this shit every day or live to do it.

Interestingly, Robert had metal items to recycle from his work as a mechanic, yet earlier in the conversation when asked about stealing metal from work, he had emphatically explained, "My company, they'll fire you for that shit." When I asked where the metal came from, he began to justify how what he had taken was not theft. It seemed that despite overcoming his difficult childhood, he continued to steal to meet his immediate needs.

Similar to Robert, a few metal thieves explained that they stole to supplement their regular income and to pay bills. One of them was Michael, who said, "I did use a lot of it just for bills and such." In at least two cases, metal thieves adjusted how much they stole and their frequency of theft based on the bills that were due. For example, Dustin discussed how he and his metal theft partner would actually "set a goal every day, each day it was different depending on if we had bills to pay." Moreover, Leo explained that he stole metal and used the profits to help his family and friends pay their bills and meet needs. He even indicated that he would steal more when his family and friends were in need of money:

I paid bills with it; I helped my family and friends out, you know. Oh, my [sibling] needs a $1000 for this, $500 hundred for this, you know, for [their] kids and stuff, you know, I would give it to [my sibling] and just do another job to replace that money.

Obtaining money to pay bills and general living expenses for themselves, and on occasion for others, was the most commonly mentioned use of the profits from theft. The specific reasons varied from unemployment, large bills coming due, and having no food, to a family member in need. Regardless of when or how a perceived financial need presented itself, thieves often stole to meet that need.

Drugs, Alcohol, and Tobacco
As described previously, drug usage was only mentioned as a personal factor related to metal theft in about 30% of the cases in the present study. Several thieves openly admitted to using drugs; however, only a

few identified addictions as the salient factor in their thefts. For example, Matt described why he steals metal: "Mainly to buy drugs. It was just to get that money to party, to get that drug." In the same way, Chris said, "I am doing it so I can go and get me some Spice (synthetic marijuana) ... Just to support my little cigarettes and stuff, just enough to get me a little bag of Spice, to make it through the day." What is more, Dustin described how his drug usage led him to drift from legal scrapping to theft:

> *Dustin*: I started scrapping [legally] in 2006, then as I started using drugs, I started taking things. I used cocaine in 2006 to current; I got ... started on painkillers, and that went on to heroin.
> *Interviewer*: Were the drugs driving you to commit crimes to get more?
> *Dustin*: Yes, I was committing crimes to use.
> *Interviewer*: How often did you scrap?
> *Dustin*: I scrapped every day because I needed money every day for drugs and you can never have enough because there were some days that you would only make $20–$30, and most of the time I split that in half with my partner.

While Matt, Chris, and Dustin discussed how they were driven to metal theft as a means of satisfying a drug addiction, the remaining metal thieves who discussed drugs did not identify addiction as the primary motivator for theft. In other words, very few metal thieves stole principally to feed a drug habit; instead, many metal thieves who also used drugs did so in a recreational manner.

Many thieves mentioned using the proceeds from metal theft to purchase alcohol or tobacco. For instance, Daniel, who explained he suffered from bipolar disorder, put it like this:

> *Daniel*: I got my money for the day, get me a pack of cigarettes and something else ... and you know (trails off and dances a little).
> *Interviewer*: And what?
> *Daniel*: I only scrapped metal cause it paid for my cigarettes and a little bit of bulldog (alcohol) and I don't do dope. I ain't been doing dope out there, but I've been on them cigarettes and stuff.

Michael explained, "In the beginning, I remember just being happy with making extra beer money."

Drug usage does play a factor in metal theft, but it does not appear to be a significant factor within the present study. Nearly two-thirds of the metal thieves interviewed said drugs were not a factor whatsoever in their thefts. Of the remaining third who discussed drug usage, only a small portion discussed drugs as the motivation for them to steal metal. Rather, when drugs or alcohol and tobacco products are mentioned, they are usually addressed in a casual manner, similar to the way society may discuss looking forward to having a drink on Friday after a long week of work.

Play Money

Perhaps the rarest use for the money earned from metal theft was what some described as "play money" or money that could be used for anything; typically, something the thief considered a fun luxury and was not designated to a particular expense. Jessica said she used her profit for "both" living money and playing around money. Leo described his excess funds after paying expenses and helping the family as "being able to live comfortably for a month," whereas, Eric, who was employed full-time while he stole, described his motivation this way:

> *Interviewer:* Did you do this because you had to have the money or was this just kind of extra play money?
> *Eric:* Extra play money.
> *Interviewer:* So this was just extra money for extra things?
> *Eric:* Yeah, it was extra money, and really I spent it on my kids, my [children] that's what I did, I gave them things that they might not be able to get, that I wouldn't be able to justify buying. You know, I do okay, and I live decent, and my kids get plenty, but you know, you would buy the stupid things that you would never buy on your own income.

Eric's partner, Michael, made similar statements:

> Can you imagine getting paid 75k a year and then getting handed an extra $200–500 bucks a week in cash? I didn't have a worry in the world. I didn't flaunt my money by any means, but every Disney video that came out and every new toy my [children] wanted, I bought it.

John was less specific, but conveyed that the money he earned was "just having spare money, basically because we enjoyed the money… [for] play money, whatever I needed it for."

Profit Conclusion

Only, three metal thieves directly recognized that their addiction to drugs was the motivating factor behind the metal theft. Most metal thieves usually listed multiple uses for their profits from metal theft. While some expenditures are emphasized more—such as living expenses, bills, alcohol, cigarettes, or providing play money—these disbursements were often not exclusive categories. In fact, most metal thieves mentioned several ways they used the profits from their crimes, making comments similar to David describing how he uses his metal theft profits: "Gas, cigarettes, you know my own personal needs ... just stuff like that. Alcohol."

7.4   THE PRICE-THEFT HYPOTHESIS AS A MOTIVATOR

Currently, the majority of research on the topic of metal theft is evaluations of the connection between the price of metal (on a commodity market) and rates of theft. A plethora of government agencies, non-government organizations, and researchers has examined the connection (in particular between copper and theft rates). The price-theft hypothesis has been statistically validated by several empirical studies and across local, national, and international areas (see Posick, Rocque, Whiteacre, & Mazeika, 2012; Sidebottom, Ashby, & Johnson, 2014; Sidebottom, Belur, Bowers, Tompson, & Johnson, 2011). This correlation remains high despite significant monthly fluctuations in the value of the metal. However, the question that remains is, what are metal thieves' perceptions of the effect that the selling price for metal has on their activities?

The metal thieves interviewed for the present study were relatively split on how the price of metal affects their behavior. About one-third of those who discussed the price of metals indicated that they do not pay attention or care if the price of metal is high or low. Rather, they only stole when they had a financial need, and their frequency or volume of theft was not dependent on the price they could receive at the recycling center. Some thieves, like Leo, who sold mass amounts of stolen copper valued between $2000 and $6000 on a regular basis casually said, "It didn't really affect me when I went [to steal].... You know, it fluctuates, it goes down 10 to 20 cents; it will go up 10 to 20 cents, so it's really not a big, big difference."

While Leo can calmly discuss the fluctuating price of metal and not be concerned with the price disparity of a few cents a pound, other thieves disagreed. Another third of the thieves unequivocally discussed how prices affected their motivation for theft. For example, David said, "Absolutely, you care, because it's hard work, there's nothing easy about, if you know what I mean!" David's comments are in line with the way some other thieves considered metal theft, as work. Many of the thieves who stated that price influenced them talked about stealing and the price of metal in much the same way as general society might discuss having respectable jobs and getting a yearly raise. Therefore, it makes sense that these thieves would be affected by the price of metals, especially since they may subconsciously consider metal theft a job.

A few other thieves discussed how the price motivated them to steal in greater volume and frequency compared to what they typically took. Zach, for example, explained this rise in relation to his activities with his friends, who are also metal thieves, saying, "There's a base [level], but [we] kick into overdrive. [We] take a lot more risks for it." Zach's comments are echoed by several scrappers (non-thieves) who discussed how higher prices motivated them to look for metal. Based on these conversations, it appears that metal thieves may put forth a fairly consistent effort to locate and collect metal. However, when the price of metal increases, thieves' motivations increase, so their income may also increase. Some thieves also suggested that their discretion is lower and they were more willing to take risks when the prices were higher.

The rest of the metal thieves gave conflicting messages about how the price motivated them to steal. In other words, they were often aware of the changes in price and discussed how it would affect their behavior; however, a few moments later, they often contradicted their statements. Consider a conversation with Dustin on how the price influenced how much he stole: "Yes, we would keep track of prices … but we never held on to anything more than two days. We did the same every day … go hard or go home!" While Dustin claimed to keep track of the prices, his admitted behavior does not demonstrate that he acted with this knowledge. This discrepancy was common with many metal thieves. Many discussed the price of metal, but few expressed identifiable changes in behavior that coincided with this influence.

It is hard to say with certainty how the price affects the motivations of metal thieves. Clearly, some in the present study identified the price as influencing their actions; others said it did not; and still others made positive comments but did not appear to follow through, based on what they had communicated. Overall, these findings are inconclusive. Clearly, there is an effect on a certain section of metal thieves, but the magnitude of the effect is unknown. Moreover, the effect may not be as apparent with seasoned metal thieves, which majority of the thieves in the present study were. Specifically, it is unclear from this study how the increased price might attract new thieves and thus account for a portion of the increases seen in other studies.

7.5   SUMMARY

This chapter began by identifying a definition for metal theft as "the theft of item(s) for the value of the constituent metals" (Whiteacre et al., 2014). Next, it examined the characteristics of metal thieves, finding that the highest proportion of metal thieves were 30-something white males, with some college experience, currently employed, and had a past work involvement in a field related to working with metals.

Next, the motivations of thieves were examined. Here several motivations were addressed, including an examination of criminal drift from legal scrapping into metal theft and criminality. Findings indicate that most metal thieves were legal scrappers at some point before they began stealing. Other motivations included an enjoyment of the criminal activity. Some of the metal thieves enjoyed the rush they received while committing crimes related to metal theft and the theft itself. Enjoying the offense was not true of all thieves; some expressed remorse and a feeling of being trapped with no other options, while others were ambivalent in their criminal enjoyment or shame.

Following the discussion of motivations was an exploration of how the profits from metal theft were used, revealing that general expenses and bills were the most common use. These profits were not only used for the thief to live, but occasionally included gifts to friends and relatives. Drugs, alcohol, and tobacco are another motivational factor for metal theft, although not to the degree commonly identified by previous government reports and practitioner publications. While three of the metal thieves in this study directly related their criminal activity to drug addiction, most mentioned drugs and alcohol as something they would buy with their proceeds in much the same way as general society might celebrate a hard day's work at the local bar.

Finally, an examination of how the price of metal motivates thieves found that they are evenly split. Nearly one-third said a higher price of metal at the scrap yard would cause them to steal more; one-third said it had no impact; the last third gave conflicting signals, verbally indicating it had an impact, yet providing contradictory details. While this study is unable to corroborate the price-theft hypothesis, interesting questions are raised as to the effect it may have on criminal behavior.

REFERENCES

Beccaria, C. (1986). *On crimes and punishment* (D. Young, Trans.). Indianapolis, IN: Hackett Publishing Company. Original work published 1764.

Ferrell, J. (2006). *Empire of scrounge: Inside the urban underground of dumpster diving, trash picking and street scavenging.* New York: New York University Press.

Matza, D. (1964). *Delinquency and drift.* Hoboken, NJ: John Wiley & Sons.

Posick, C., Rocque, M., Whiteacre, K., & Mazeika, D. (2012). Examining metal theft in context: An opportunity theory approach. *Justice Research and Policy, 14*(2), 79–102.

Sidebottom, A., Ashby, M., & Johnson, S. D. (2014). Copper cable theft revisiting the price–theft hypothesis. *Journal of Research in Crime and Delinquency, 51*(5), 684–700.

Sidebottom, A., Belur, J., Bowers, K., Tompson, L., & Johnson, S. D. (2011). Theft in price-volatile markets: On the relationship between copper price and copper theft. *Journal of Research in Crime and Delinquency, 48*(3), 396–418.

U.S. Census Bureau. (2015). Educational attainment. Retrieved March 2015, from https://www.census.gov/hhes/socdemo/education/data/cps/2014/tables.html

Whiteacre, K. (2014, November 4). (B. Stickle, Interviewer).

Whiteacre, K., Terheide, D., & Biggs, B. (2014). *Research brief: Metal thefts in Indianapolis October 1, 2011–September 30, 2013.* Indianapolis: University of Indianapolis Community Research Center.

Zimring, C. A. (2009). *Cash for your trash: Scrap recycling in America.* Piscataway, NJ: Rutgers University Press.

CHAPTER 8

Techniques and Methods of Metal Thieves

This chapter examines the factors thieves consider when identifying, locating, acquiring, and selling stolen metal. The chapter also provides a general discussion of places, including how places influence theft, as well as some of the techniques used to avoid detection. Three different stages are involved in metal theft: identifying a place with metal, gaining access and taking the metal, and selling the metal to another so that the profit can be enjoyed. While these three stages are not entirely unique from other types of theft, the activities which occur during each stage are necessary to study in order to identify the stages of the theft that can be interrupted. Prior to discussing the findings of this study, however, this chapter examines the current body of research on metal theft as a crime of place.

8.1 SPATIAL ANALYSIS

As discussed previously, metal theft is a crime affecting the built environment (Bennett, 2008). The built environment involves spaces and places created or modified for human activity and spans parks, churches, farms, industrial buildings, homes, and any other space adapted by people. In most situations, the built environment includes the use of metals such as steel, copper, iron, aluminum, and lead, which implies that with the current high value of metals, the built environment is "an asset under attack"

© The Author(s) 2017
B.F. Stickle, *Metal Scrappers and Thieves*,
DOI 10.1007/978-3-319-57502-5_8

(Bennett, 2008, p. 176). While some places include metals that are harder to access (e.g., buried copper pipes and wires in a field), other places possess metals which are easily accessed (e.g., aluminum siding, bronze grave markers in cemeteries). Moreover, some places possess greater quantities of valuable base metals in a small location (e.g., copper wiring at industrial facilities), whereas others possess fewer supplies of valuable metals spread across a sizable area (e.g., farms). Further, lack of guardianship of some locations makes theft at some places easier. Finally, the location of recycling centers (a necessary component to metal theft) may also influence the rates of crime. These and other factors will be examined in this section.

8.1.1   Place and Ease of Access

The ease of acquiring metal at certain places appears to be a significant factor in metal theft as is demonstrated in the types of places that metal thieves target. For example, agricultural businesses and farms are often targeted due to the ease of acquiring metal (Washington State Department of Agriculture, 2015). Thieves can raid barns filled with metal, haul off old farm equipment, and remove electrical and irrigation equipment with little effort and a reduced likelihood of being discovered. In fact, California's Agricultural Crime Technology Information and Operations Network (ACTION) project identified a 400% increase in metal thefts on farms in 13 counties in California between 2005 and 2006 (Souza, 2015).

Other places also provide significant ease of access to valuable metals within the built environment. For example, catalytic converter theft requires only a few moments, a saw, and a victim vehicle with high ground clearance for thieves to quickly cut the expensive exhaust system from a vehicle and turn an $80 investment into a $200 profit. Places such as parking lots (especially long-term parking lots), vehicle repair shops, and car sales lots are likely places thieves would target, as they are rich with opportunities and often have limited guardianship.

Likewise, construction sites are prime places that may attract or generate crimes. Construction sites frequently have large quantities of metal items at each stage of development (e.g., metal tools, copper pipes, and electrical wire). A construction worker may purposely steal metal from his employer while on the job site. Moreover, metal thieves who are not employed at the location may seek construction sites out to capitalize on the large amounts of metal.

Finally, some public places within the built environment are especially vulnerable to metal theft. Examples include alleyways, parks, and cemeteries.

Each of these places provides an easy opportunity for crime, as metal is often abundant and guardianship is low, and the metal is often easily accessed. As an illustration, consider the aluminum siding on a garage or home that adjoins an alley. Since metal thieves may search along alleyways for readily available metal, the aluminum siding may be attractive. Moreover, cemeteries and parks often have poorly secured bronze decorative art or plaques. In these cases, circumstances require only that a thief visit an ordinary public place, pick up a metal item, and abscond.

8.1.2   Quantity and Quality of Metal

The rational choice perspective (Cornish & Clarke, 2014) postulates that crimes are more likely to transpire when there is a higher likelihood of financial gain. There is limited direct research examining this issue in relation to the crime of metal theft. However, anecdotal findings indicate that metal theft often occurs in large quantities and that metals of high value (quality) are frequently targeted. For example, copper is the most stolen metal (Kudla, 2012; NICB, 2014; Whiteacre, Medler, Rhoton, & Howes, 2008; Whiteacre, Terheide, & Biggs, 2014), likely because it provides more income per pound than most other metals. It is also likely that places with high quantities of quality metals are targeted for crimes more often than places that do not possess these characteristics. For example, older homes, which tend to have copper pipes, as opposed to plastic pipes, may be a place targeted by thieves.

8.1.3   Lack of Guardianship

Routine activities theory (Cohen & Felson, 1979) postulates that crimes are more likely to occur in places of low guardianship where the crime is less likely to be discovered. A lack of capable guardianship can be identified in many of the places that often suffer metal theft; these include railroads (Ashby, Bowers, Borrion, & Fujiuama, 2014; Posick, Rocque, Whiteacre, & Mazeika, 2012; Robb, Coupe, & Ariel, 2015; Sidebottom, Ashby, & Johnson, 2014; Sidebottom, Belur, Bowers, Tompson, & Johnson, 2011), farms (see discussion above), and industrial and construction locations (Berg & Hinze, 2005; Boba & Roberto Santos, 2007; Clarke & Goldstein, 2014). However, due to the recent US economic downturn and increased home foreclosures, abandoned buildings have also received increased attention as crime generators and attractors (Stucky, Ottensmann, & Payton, 2012; Zhang & McCord, 2014). As with other

types of crime, abandoned homes and buildings appear to have a high correlation with metal theft.

For example, Posick et al. (2012) examined metal related burglaries
and non-metal related burglaries in Rochester, New York, between April
2008 and July 2010. They found that burglaries related to metal thefts
were more likely to occur when neighbors were further away (average of
138 meters) than for burglaries that were not related to metal (average of
44 meters). The study also demonstrated that 54% of all metal related burglaries occurred in vacant buildings, compared to only 3% of non-metal
related burglaries. The authors performed a logistic regression of the data
and found that the relationship between vacant buildings and metal burglary was "very substantial ... increasing the log odds of metal being stolen by 44.68" (p. 94). This data appears to support the idea that lack of
a capable guardian is a substantial factor for metal thieves when choosing
places in which to commit theft.

8.1.4   The Role of Recycling Centers

The profit for the thief involved in metal theft requires a specialized buyer.
"Scrap buyers provide the necessary link for creating profit from scrap metal
theft. The scrap metal theft problem is driven entirely by the ability to sell
stolen goods to recyclers, and often these recyclers facilitate crime" (Kooi,
2010, p. 7). Recycling centers typically take the form of an established business that buys metal from the public. These recycling centers continue to
separate and process metals, and commonly resell metals to larger and larger
buyers as the metal re-enters the recovery cycle. Unfortunately, it is challenging for recycling centers to differentiate stolen metal from legally obtained
metal. Research has demonstrated that certian types of businesses may have
an effect on crime trends in their immediate surroundings (McCord &
Tewksbury, 2013). Whiteacre and Howes (2009) postulated:

> By unknowingly (or sometimes knowingly) purchasing stolen items, recy-
> cling centers may facilitate the disposal of stolen goods, thus increasing the
> theft of those items. The presence of scrap yards, therefore, might play a role
> in the increasing metal thefts in the area. (p. 2)

To examine this theory, Whiteacre and Howes (2009) utilized NICB claims
data from January 2006 to November 2008 for 51 cities across the United
States that had experienced high rates (30 or more) of metal theft claims.

The data was then compared with the rate of publicized recycling centers per 100,000 residences in each town. The researchers discovered a "strong, positive and significant relationship with the number of scrap yards" (p. 7). The findings also indicated that metal theft had a positive correlation with burglary rates in each city, but that the relationship was not as strong as the number of recycling centers. Whiteacre and Howes examined the data through a linear multiple regression analysis and found that recycling center rates accounted for 52% (r-square 0.527) of the variance in metal theft.

Despite these findings, Whiteacre and Howes (2009) discuss several important caveats. First, the data acquired for this study, namely the NICB data, significantly underrepresents metal theft. Second, their findings do not indicate causation, only a positive correlation. Lastly, Whiteacre and Howes caution against interpreting the results to suggest that recycling centers are crime facilitators. They cite Sutton (1995), who examined the role of pawn shops with non-metal theft and stated, "The question remains as to whether the existence of a market for stolen goods is merely a downstream consequence of crime, or whether it feeds back to provide motivation for thefts and to influence what is actually stolen" (p. 400). The link between recycling centers and metal theft is apparent. Thieves need to sell their stolen material, and recycling centers either knowingly or unknowing provide the outlet for such sales. However, the role of recycling centers in metal theft and the extent to which they influence thieves remain unknown.

8.1.5   Spatial Analysis Conclusion

The findings from the available empirical research tend to support the hypothesis that metal theft is more likely to occur in particular places when high quantity and high quality are combined with decreased guardianship. Moreover, the likelihood of metal theft is enhanced when acquiring metal can be accomplished without significant effort. Further, it appears as though the factors that are attractive to metal thieves—ease of access, quantity and quality of metal, and lack of guardianship—often coalesce in specific places and create high probabilities of crime. For example, abandoned buildings provide easy access, a large supply of quality metal (e.g., copper), and a shield between the outside world and the criminals (reduced guardianship).

Similarly, churches are other places that are usually not occupied (low guardianship), with easily accessible air conditioners, which contain a significant amount of valuable copper. Thus they are also frequently victims of metal theft. In fact, Whiteacre et al. (2008) discovered that churches in

Indianapolis, Indiana, were disproportionately victims of metal theft compared to other types of structures. Moreover, the Ecclesiastical Insurance
Group (2014) of the United Kingdom, providers of insurance to religious
institutions, reported more than 11,000 insurance claims related to metal
theft between 2007 and 2013. It would appear that when places possess the
qualities discussed in this section and are near a recycling center, the instances
of metal theft may rise. Unfortunately, research also demonstrates that once
a place has been the victim of metal theft, it is likely to suffer a repeat victimization (Ashby et al., 2014; Whiteacre, Terheide, & Biggs, 2015).

8.2   THE PLACE

As demonstrated, the built environment significantly factors into the place
of metal theft. Items stolen are often physical structures or parts of a structure (i.e., down spouts, copper pipes) and frequently only have extrinsic
value to the thief. Places that attract metal thieves tend to have three common characteristics. They have a tendency to have reduced guardianship
(e.g., abandoned buildings), have a high quantity of metal (e.g., electrical
substation), and a high quality of metals (e.g., copper). Metal thieves are
frequently enticed to places where these three factors coalesce. The process by which metal thieves identify these places is examined here. Two
primary activities occur at this stage. The first is identifying targets and the
second is casing and planning the theft.

8.2.1   Identifying Targets

In this study, the most common category of places targeted by thieves was
abandoned structures, which included abandoned residential homes, as well
as abandoned business and industrial sites. These places often contained a
significant amount of quality metals (copper) and had very limited guardianship. Both urban and rural metal thieves often knew which businesses
and homes were abandoned because they were familiar with the neighborhood. For example, Matt described how he and his partners would identify
abandoned homes: "You just knew they were abandoned, no movement for
weeks or months. Everybody in the [neighborhood] knew what was abandoned." Similarly, Jessica who stole in a rural environment explained how
her group of metal thieves operated: "We knew when a house was coming
empty. 'Cause like I say, we were such a small town; everyone knew everyone, and we knew someone who had lived there and just moved." Matt
and Jessica's statements illustrate what many metal thieves indicated; they

tend to operate within an area where they were familiar and comfortable. Moreover, their knowledge of the area provided the information helpful to identify abandoned homes—sometimes, as in Jessica's case, immediately after someone moved out and before a new tenant moved in.

Metal Thief Profile

Name:	Jessica
Taxonomy:	Metal Thief
Location:	Southeast—Rural
Age:	35
Race:	White
Education:	Associates
Employment:	Full-Time—Nursing Home
Experience:	After my conversation with Chris (see Chap. 7), I noticed Jessica and some other persons looking through a raised building. When I stopped to say hi, Jessica was eager to talk while her boyfriend and his crew separated useful wood and brick from the building that had recently been torn down. With her five-year-old playing in the construction site nearby, we sat on five-gallon buckets and talked about her experiences. She discussed how her former boyfriend got her hooked on stealing metal after she had seen a news report of a thief who had stolen thousands of dollars' worth of metal, and she became interested. She described how she previously lived in a small town and after she got off work from her nursing home job she would go out stealing metal and use the money for drugs. She began by simply being the driver but progressed into an active participant. Jessica seemed happy to talk with me, and told me many details of her crimes and explained the techniques of locating abandoned buildings and homes. She also described how they would go out of city and county jurisdictions to scrap yards to avoid "the law." Jessica explained that she was eventually in prison for five years on a non-metal-theft-related charge and since then has lived a clean life with a new boyfriend.

While many metal thieves had an intimate knowledge of the area they stole from, some did not. If a metal thief did not have a well-developed understanding of the area and was unable to identify abandoned buildings, that information would come from somewhere else. Several metal thieves obtained information on abandoned buildings from others who were not involved in metal theft but were familiar with an area that contained abandoned buildings. Chris, for example, was not originally from the area in which he was stealing and therefore made contact with someone who purportedly owned abandoned buildings and allowed him to strip them. He explained, "I got a guy, here in [the city]. He's got 12 vacant homes, and they are getting ready to tear them down. They are all vacant, abandoned … so he just lets me go through and [strip them]." However, after further discussion, Chris explained that his source did not really own the buildings but was only "helping me out with the location."

Leo's technique for locating abandoned properties was unique to the present study. Given that he disclosed earning a profit of nearly $250,000 by stripping 10 abandoned buildings within three years, his technique is important to include. Leo explained how he would,

> go to Google and put in abandoned buildings and [the state] and a bunch of places would pop up. Then you can go on YouTube and watch videos because people will go inside the buildings and make videos, and you can go and look at the video to see if you can see any copper in the building before we go.

However, this is not how Leo began locating abandoned buildings. Originally, Leo had knowledge of a few abandoned buildings near his home and work. When those buildings were depleted of copper, Leo began to search for other abandoned buildings online and would travel up to 50 miles away to steal metal.

The second most common kind of places preferred were those that had a combination of place characteristics and air conditioners. For example, metal thieves more likely targeted air conditioners that were located in areas with reduced visibility from the public, that were in multiples, and that were large units. Most metal thieves frequently seek after air conditioners while others exclusively target them. Matt, who worked as a HVAC technician, described how he would target the places he was working to steal, saying, "We would [install] a unit, [then] scope out the neighborhood and snatch theirs. We would snatch three or four in a day, easy." On occasion, Matt even returned to where he knew he had installed large units, which were worth more money, and steal those. "We actually put those three up, and about two weeks later, we were back there snatching them." Similarly, Dustin often targeted places with multiple air conditioner units, like apartment complexes where he had worked as general maintenance personnel in the past, and described how the targeted place "had to be concealed somehow by trees or behind a building." However, he would never steal from any place where he was actively employed, saying "Don't scrap where you eat." Moreover, if the apartment complex was not physically conducive to theft (e.g., lack of concealment, too many people), Dustin would identify shopping centers that had multiple air conditioner units located in less frequented areas, such as units lined up behind the stores.

Businesses that had significant amounts of metal on hand were also frequent targets. The thefts nearly always occurred after a business closed, the employees were gone, and once it was dark. The most common businesses stolen from were vehicle junkyards and auto body shops that had metals stored outside. These places also had large amounts of valuable materials such as aluminum wheels, catalytic converters, and batteries. Other businesses were also targeted. For example, Zach relayed a story of being arrested while he and his metal theft partner "threw street signs over the fence into his truck" from a company which manufactured street signs. While some businesses were frequent victims, it seems any company that had substantial amounts of metal on hand was a likely target.

Finally, farms and other rural areas, such as forests, were common places for metal thieves to focus. These places tended to have large quantities of metal available in the form of equipment, vehicles, or other metal structures, with little to no guardianship (e.g., abandoned buildings, old trash dumps, rarely used equipment). A thief rarely wandered around looking for metal at these places; instead, it was common for the thief to have had a previous legitimate reason to be on the property and to have identified something to come back and take. For instance, David described the logging equipment he was aware of:

> [At] logging operations, they left the old equipment sitting there, there's a lot of old equipment sitting out, it's been sitting there forever. I am talking like huge, bulldozers or something. Bulldozers or booms, you know,

the big logging trucks for dragging logs up the canyons, you know. I could probably take you to a 10-mile radius right now; I could probably take you to [three or four hundred] tons of nothing but leftover logging equipment just sitting there looking at you.

While many of the metal thieves indicated that finding places from which to steal metal was easy, a few discussed how difficult it was to find places with the same volume of metal as when they began stealing years ago. The struggle to find metal coincides with comments by many scrappers who lamented the difficulty in locating metal, as interest in metal has increased. Several metal thieves discussed how competition is getting tight and suitable places with lots of metal are becoming scarce. Leo explained it this way:

> Leo: It's pretty simple to do, but like I said, toward the end it was getting a little harder to do, not going into the place and scouting, but actually finding a place. That's actually, I would have to say, that's the hardest part is actually finding a place.
>
> Interviewer: Okay, and why do you think it's getting more difficult?
>
> Leo: Because there's so many people doing it, so you'll find a place, and just everything will be gone, once a place goes abandoned, and people hear about it, it's like a free-for-all, you know.

Regardless of how easy or difficult it is to identify and locate a possible place to steal metal, that is only the first part of this stage.

8.2.2   Casing and Planning

After the identification of a place, most thieves cased the location and planned a method to accomplish the theft. Sometimes casing was as simple as checking to ensure that the residents were asleep before stealing an air conditioner, or looking for dogs and cameras. Other times, casing the place was extensive and included planning methods to effect the theft.

Typically, when entering a structure to steal metal, increased surveillance and planning occurred. For instance, planning a burglary to acquire metal often involved entering the building before the actual theft to ensure there was metal present. As Chris explained, he would "just go in and find out ... if there is wire [and] shit in there." Chris was not alone in the tactic, as several thieves discussed similar activities. While talking with Jessica,

I pointed out an abandoned building in the distance and asked how she would approach it. She disclosed:

> So what we would do is keep an eye on it for two or three days, to make sure no one went in and out. Scoped out to make sure there were no security cameras. And then, say like the neighborhoods, we would watch the neighbors to see how late they stayed up, how much company they had, shit like that. So that way you could get a time frame to go in there to strip it out. It's a process.

The only common exception to casing and planning a burglary seemed to be when a thief already had a working knowledge of the building. This could be a home where the thief had been before, such as a friend's house. In those situations, thieves were more likely to enter without casing the building.

Metal thefts that occurred outdoors, such as stealing logging equipment, taking air conditioners, or removing catalytic converters from vehicles, involved much less planning. The majority of these situations included only a cursory glance to ensure that no one was watching and often occurred a night. The reduced planning occurred for several reasons, one of which is the speed with which many of these types of crimes can happen. For example, it takes only a few moments to remove an air conditioner.

Another reason for reduced casing and planning is the less severe nature of the possible charges. Metal theft involving abandoned buildings could result in charges of burglary, which is a felony with likely jail time, compared to trespassing and theft. Metal thieves understood the gravity of burglary charges and tended to take entering abandoned buildings more seriously. As a result, such burglaries tended to be well planned and the places cased before committing the theft, whereas theft occurring outside buildings tended to be less planned.

8.3   THE THEFT

Once a place containing a likely supply of metal has been identified and, if necessary, casing and planning have been completed, the second stage is stealing the metal. The methods and techniques used to accomplish the theft are largely dependent on the place, type, amount of metal, and often on the skill of the thief. These factors play a role not only in the techniques (e.g., tools, tactics, transportation) that a metal thief employs, but also in the methods (e.g., time of the theft, frequency, repeated thefts). This section will examine these issues, identifying the problems related to theft techniques and methods.

8.3.1   Frequency and Time of Day

The rate of theft varies, primarily based on the financial need of the thief. Some thieves, like Chris, stole small amounts of metal each day so they could live, but the majority steal once or twice a week. Moreover, a few thieves, like Leo, steal less frequently, about once a month. In nearly all circumstances, thieves described taking metal to meet a financial need. Leo even described the large abandoned buildings he stole from as banks, saying, "Once you're starting to run low on money you know just go back to that building and get another $4000 or $5000." Explaining the frequency of his theft, Dustin said, "At the end, I was scrapping every day, both legal and illegal."

Metal thieves are evenly split on when they stole metal, night versus day. To a great extent, the time of day when thieves stole was dependent on the place. For example, if thieves were targeting a business with an outdoor supply of metal, they would always go at night after the establishment had closed. Conversely, most thieves who stole from rural areas, such as farms or forests, often went during the day. Stealing during the day in these places was important not only so they could see what they were doing, but also because it was less suspicious. John, for example, discussed stealing items from a farm during the day, saying, "We did it during the day [and] if anybody asked us any questions we just told them that it was our tobacco frames, or this and that, or told them we was getting it for someone else, or whatever."

Air conditioner thefts present an unusual paradox concerning the time of day. For example, Matt would often steal air conditioners right after work (in daylight), while he was still in his company uniform and driving the company truck, thus reducing the questions someone might ask if he were asked about the metal. Conversely, Dustin, who lacked a strong cover for his activities, said about scrapping and metal theft, "all day if legal, all night if illegal." Specifically disclosing when he stole air conditioners, he said, "No [we] did it at night if it (the residence) was occupied, when they were asleep." Working at night provided Dustin with the cover of darkness and less traffic so that he could steal air conditioners from

Image 8.1 An air conditioner being disassembled on the roadway around the corner from a scrap yard. The Subsistence Scrapper claimed that a construction crew gave it to him

occupied homes and apartments. Overall, those who stole air conditioners were split on when they stole; some during the day and others at night.

Most burglary-related metal thefts also occurred during the day, primarily so that thieves could see what they were doing and so that it would not be as visible to anyone who might pass by the structure that there were people inside. Leo described his logic, which encompasses much of what the others described about burglary during the day, saying, "Because you can see [in the day] and at night if you're in there, and you have to use a flashlight, and if people see the light inside the building, people can call the cops."

8.3.2   Repeat Victimization

A study of metal theft in the UK railway system by Ashby, Bowers, Borrion, and Fujiyama (2014) found that the incidences of repeated victimization involving metal theft were more frequent and longer-lasting than for most other types of crimes. The present study's findings support the conclusion that thieves often return to steal more metal and may do so for long periods. For example, nearly all the metal thieves interviewed described incidents where they would return to gather more metal. Of those interviewed for the present study, repeat thefts from the same place typically occurred for one of two reasons. First, the metal may be available on a repeated basis, which can happen because the metal item was replaced, such as a replacement air conditioner. This established a prosperous place for the theft, to return looking for the replaced metal.

The second reason for repeat victimization is because there may be so much metal the thieves cannot steal it all at once. An abundant supply was often the case when thieves burglarized a large or older building. Even a smaller home would often require several thieves hours—if not several days—to remove all the copper pipes and wires. Therefore, some thieves would take what they could in one night and wait to come back at another time. Each time they returned they would take more until the supply was depleted. Moreover, in situations where there is a significant amount of material the work is challenging. Leo explained, "It takes a lot of work and a lot of energy, at the end of the day I'm so tired it's not even funny. My body hurts tremendously." Due to the physical work that is involved in stripping a large structure and the income that is produced (e.g., several thousand dollars), Leo usually only stole metal once a month and would then return to the same place until all the metal was gone.

While most thieves did not make thousands of dollars from a single instance of theft, they often made several hundred. The amount of profit also factors into how often thieves return to the same place. Many thieves, as already mentioned, are motivated to commit theft when they need funds. Consequently, when they have enough money, they are less likely to steal. When a need arises, however, thieves often return to areas that have the preferred criteria for metal theft: a secluded area with limited guardianship and with the easy availability of high quality and quantities of metal. In other words, some places are repeatedly frequented by metal thieves due to their physical characteristics.

Lastly, some thieves who have collaborated with each other might split up, and then return individually to steal metal. John's partner, for example, returned several times without him to steal new farm implements. It is possible that when a significant amount of metal is found, some of the group may return individually before the others. A metal theft partnership whose members begin to individually steal from the same place increases the frequency of theft at the same place.

8.3.3   Tactics, Tools, and Transportation

Metal thieves employed in an industry that deals with metal are afforded operational cover for their thieving activities and have an excuse to be in possession of large amounts of metal. Dustin, for example, used his employment to conceal his nefarious activities. He explained that if someone approached him and his partner while either of them were scoping out an area or actively stealing an air conditioner, they "acted like we were maintenance people," which indeed they were, but at another location.

However, beyond using work as a cover or for locating metal, much of the technical knowledge necessary to quickly complete metal theft was also gained while at work. For example, Jessica said, "I worked at the scrap yard," when asked how she learned to identify which metals were more valuable. Moreover, Matt's training as a HVAC technician and licensed journeyman gave him the knowledge to steal an air conditioner in under four minutes. Matt also used the tools from work to complete the theft, saying, "Freon gauges and valves ... I had it all, the whole nine."

Not everyone had advanced tools like gauges and valves from work. However, most metal thieves had a basic assortment of tools to complete the theft, similar to what Leo described, "Bolt cutters for one, bolt cutters can cut small copper pipe and wire, and it will cut through the cable. Hacksaws, screwdrivers—flat head and Philips—them little pipe cutters, lots, and lots of duct tape." While stealing metal, it was standard for a metal thief to wear gloves for several reasons; as Dustin put it, "[To reduce] fingerprints and also ... so [I] didn't get cut as easy."

Once the metal is stolen, it has to be removed from the place and taken to a recycling center, which was nearly always done by vehicle. There was no consistency in vehicles, and everything from shopping carts to cars, vans, and work trucks were discussed. The type of vehicle usually depended on what was available and what was sufficient to contain the metal. For example, pulling large farm implements required a truck, while an air conditioner will fit in the back of a medium-sized van. Interestingly, loading the vehicle and driving to the recycling center were often spoken of as the tensest times. Eric explained how he would regularly transport over a ton of metal in a Ford Escort:

> I was always a little concerned. I would always make sure about the weight that was in the car; you could easily tell something was in there. Well, I always made sure that my tires weren't swayed real low, you know, I wasn't going to start driving and not try to be the most prepared I could be to make sure that this car was going to be fine.

Image 8.2 This thief kept his tools neatly inside this pouch. Several other tools were being used at the time and are not pictured, including a knife, magnet, pliers, and screwdrivers

Eric's primary concern was that he would break down or get a flat tire, and law enforcement would arrive. His fears were well founded, as Leo and his partner were captured with two tons of copper on the way to a recycling center after their pickup truck had a flat tire.

8.4   The Exchange

The third and last stage necessary for a metal thief is exchanging metal for money. This final stage usually occurs at a recycling center and is one of the unique aspects of metal theft. In this study, there were no circumstances when a thief kept the metal for personal use; stolen metal was always sold to a recycling center for its intrinsic value or the value of the constituent parts. With few exceptions, this stage of metal theft is unique from many other forms of theft.

Recycling centers purchase metal from the public and may therefore purchase legally obtained metals as well as stolen metals. Unfortunately, most metals do not have unique identifying marks (e.g., serial numbers) or have often been damaged or altered, which makes the stolen metal difficult to identify. Therefore, determining which metals are stolen can be difficult, especially when metals are sold in a large quantity or have been highly processed (all the different types of metals separated). Regardless, recycling centers are the primary focus of the majority of laws and regulations and even many law enforcement investigations (Burnett, Kussainov, & Hull, 2014). Many believe that recycling centers knowingly accept stolen metal or are at least complicit in the theft of stolen material (Whiteacre, 2014).

8.4.1   *Criminal Exchange*

In the present study, only one metal thief, Eric, appeared to have a positive criminal relationship with a recycling center. Eric's partner, Michael, was stealing industrial sized electrical cable (e.g., with a copper diameter of over one inch) from work. Eric would strip the protective rubber coating, melt the lead off the cable, and take approximately a ton of the copper wire to the same recycling center each Friday.

This routine went on for years before Eric and Michael were both arrested. Eric described his perceptions of what the recycling center knew, how it operated, and even how it shielded him from law enforcement:

> Eric: My perception is they knew darn good and well it was hot. Yeah, they knew; they knew I was a good customer. They took care of me. I mean I was in line at the recycling center, some days it would be backed up … and there would be some police in there or something like that just looking for catalytic converters or something. There would always be a guy from the recycling center … walk out … to let me know that there was police. They'd say, "Just sit tight, they're just looking for some stolen catalytic converters," but they would always let me know.
> Interviewer: So that's something they wouldn't do to everybody else, just picking you out that they were concerned about?
> Eric: Right, as far as I know, yeah I think I was one of the few that they did, and I was actually having cell phone contact to make sure everything was cool before I brought the stuff over and things like that.
> Interviewer: Who did you have contact with?

Image 8.3 An example of a large quantity (600 pounds) of industrial-sized electrical cable stripped of his rubber sheath and ready for sale

Eric: A couple of different workers. One guy, he was a worker, but he might have been a little bit above some of the other guys, you know. I never really knew what you would call him [his designation] ... but he would always pay me, too. There has been times too where the bigwigs up in the office, they would come down. They were fine with it. They would shake my hand and thank me for my business.

Interviewer: ... did [the recycling company] know that it was coming from [the electrical company]?

Eric: I assume they had to assume it was coming from a big [job], I mean ... these were big pieces of cable, very big pieces of wire. If they did know (where it was coming from), they never let on. They never led me to believe, I mean I went in there just to get in and get out. It was like unspoken; let's don't talk about it.

As Eric stated, he believed the recycling center knew the metal was stolen, but they did not seem to care. In fact, according to Michael, Eric's partner, the recycling center encouraged both of them by warning Eric of a police presence and even providing sports tickets and other items to entice him to return. Eventually, this recycling company was criminally investigated by federal law enforcement when the theft partnership that Eric and Michael were a part of was discovered; ultimately, however, the company avoided prosecution.

8.4.2   Surreptitious Exchanges

While it seems clear that the recycling center used by Eric knew, or at least should have known, that the metal it was purchasing was stolen, it is not always easy to identify stolen metal or metal thieves. In fact, nearly all the thieves interviewed said that they took steps to avoid detection at the yards. Many of the thieves mentioned traveling out of the county where the theft occurred to sell their stolen goods because, as Jessica revealed, "it's harder to get caught up like that." Moreover, Leo explained that taking metal to a recycling center out of the county from where it came was "a standing rule." The primary purpose was to complicate or avoid police investigation by crossing jurisdictional boundaries.

Another common tactic was, as Matt put it, "You shop around!" indicating that he would distribute parts of stolen metal across many different yards. This was the second most frequently mentioned method to sell stolen metal, and was implemented in two ways. First, a thief may separate his metals into different types and sell smaller amounts at various recycling centers. For example, after stripping a house, a thief may sell a few pounds' worth of copper pipes at several different yards throughout the day. Alternatively, thieves may sell all of their stolen metal at once, but then not return to that recycling center for several weeks or months. Both of these methods were chosen to reduce the suspicions of the recycling center. Leo described how he commonly sold several tons of stolen copper once or twice a month:

> *Leo*: Various scrap yards ... we would mix it up a little bit you know. We did take it to different scrap yards around the area.
> *Interviewer*: Why did you mix it up? Was there a particular reason?
> *Leo*: Honestly, just being cautious, you know. I mean we didn't want them to see our faces too much, you know what I mean, and then they start asking questions like, "Where are you guys getting all of this stuff?" I really don't think the scrap yards care because they're making crazy money on this stuff ... It was just for our peace of mind.

Another method thieves used to avoid detection was to alter the identity of the item or severely damage it. This method occurred when thieves damaged an item and therefore presented the recycling center with a reasonable excuse to scrap the whole object. For example, John explained how he and his partner would steal tobacco frames (a large farming implement), take a sledgehammer, "And we would take them and mess them up like they weren't good anymore." This provided a logical explanation to the inquiring recycling center, as he would tell them, "We bought brand new tobacco frames and these (the ones he brought to recycle) was kinda old and bent and all that kinda stuff, and we just want to scrap them because we [don't] have anything else to do with them."

Providing a story behind a damaged item was also frequently mentioned by recycling center employees. Many of them had entertaining stories of ill-conceived attempts to show an item had been damaged and was therefore worthless. In those circumstances, most of the recycling center employees who shared such stories with me called law enforcement. Paradoxically, this is how John was arrested. His partner began stealing tobacco frames without him and doing a poor job of damaging the structures. John explained, "He was starting to take ones that were brand new ... that don't have nothing wrong with it but maybe a sliced tire that he sliced right before he pulled in, they're going to know something's up." Scrapping what appeared to be new farm implements aroused the suspicion of the recycling center, which notified law enforcement and resulted in the arrest of both men.

An alternative technique for selling stolen metal at recycling centers was to include it in a large volume of legally obtained metal. This technique often also involved damaging or cutting up the stolen metal and then mingling it in with other metals. David, who did this often, described comingling stolen and legal metal by saying, "If you get stuff cut up and bunch [it together], it all looks like stuff cut up and mixed," adding, "and those cops aren't going to go out there and look through that." While a few thieves utilized this technique, it only worked when they already possessed an abundant supply of legally obtained metal, which most thieves did not.

However, if a thief did not have the needed volume of legal metal, he could contact a fence. In these situations, a fence is a person who purchases stolen metal and then sells it grouped together with legally obtained metal, acting as an intermediary in the sale of metal. The present study was unable to locate and interview a fence, but their existence and activities were mentioned several times by more than a few individuals. For example, Michael discussed his knowledge of a metal theft fence thus:

> We never used him, but I talked to some of the guys that did, and they said he would show up in an old, beat-up truck and put a portable scale on the ground and weigh it all and load it and pay cash right on the spot. He didn't ask questions and didn't give receipts either. We all know what the deal was there! I do know he didn't give as good [a] price as the yards did. Maybe he could be considered a middleman. I do know he gave out a business card that had a name of a recycling center on it. I don't remember that name, though.

Another example of fencing metal came from Jen, who is a Subsistence Scrapper living and scrapping in the government housing projects. Jen described how drug dealers are beginning to scrap, even stealing metal items from her, and use a fence to sell their metal:

> *Jen*: The dope dealers in my community are now selling scrap!
> *Interviewer*: Now the people who are slinging the dope and are scrapping, are they doing it (referring to collecting scrap) the same way you are doing it?

Jen: Well, like I told you, the maintenance people give me stuff. And when they see the maintenance truck give me something [the drug dealers], tell me I am in their way and they want it. They are not getting in the dumpsters and all like I do.

Interviewer: So are they bringing it down here, too?

Jen: What they were doing that day is getting somebody else who scraps and giving it to them so he could pay them.

Interviewer: So dope dealers see stuff, call somebody they know who is a scrapper, and say I will take $5 for this AC unit and the person will bring it down here and get $10.

Jen: Yes! Yes! Exactly! Right now that is what they are doing.

As the laws restricting what people are allowed to sell at scrap yards and the rules and regulations expand (e.g., a government issued photo ID, recording pictures of the metal sold), it is likely that fencing metal will increase as well. As has been seen in these two examples, the practice may already be in place and may be common knowledge among thieves. For example, when I asked Zach what to do with stolen metal, he said to "call the guys [who list their phone numbers in the papers]; they will come to you and buy anything." Zach and Jen were referring to Professional Scrappers (described in Chap. 3). According to several thieves and scrappers, some Professional Scrappers are knowingly purchasing stolen material and therefore operating as fences. Unfortunately, individuals who scrap and fence stolen metal by mixing their load of legal and illegal metals are tough to differentiate, and it is doubtful that recycling centers are aware when this activity occurs.

8.4.3   Trusted Sales

The last technique some metal thieves use to sell stolen metal items to the recycling center was to do so in the course of their regular business activities. In other words, criminals would present themselves as recycling metals legitimately gained in the course of work. Most recycling centers tended to trust these individuals and had no suspicion of their material, as they were employees of another company (e.g., electricians, plumbers). Trusted sales were accomplished primarily in two ways. The first is similar to others who intersperse stolen metals in with legal metals; the distinction is that the thief works for a place where he or she has legitimate access to metals. For example, Matt worked in the family HVAC business and described how he used this technique:

[I'd] mix the [stolen] scrap that I had in with that [legal] scrap… Sometimes after a big job, like we did a big job for [a local university] and the library … [I'd take it] (his stolen metal). 'Cause they (the yard employees) knew me. They knew, ahh that is Billy's son. I worked for [a large air conditioning installation and repair company]. But nobody really knew I was doing that. The only guys that knew were the ones I was running around [with] … that was it. Everybody else didn't… nobody know, nobody knew. They (the yard employees) did not have a clue!

A few metal thieves appear to sell stolen metal outright, without attempting to hide its origin or mix it in with other legal metal. Combining stolen metal with non-stolen metal is the second way that metal thieves sell metal in the course of their business, without skepticism from recycling center employees. Simply by working for a company that has valid and justifiable reasons to possess metal, they can pass in and out of recycling centers with stolen metal. Individuals in this category include licensed plumbers, HVAC workers, electricians, and those who were employed replacing windows, siding, and even general contractors. While conducting observation at several recycling centers, I watched individuals arrive near the end of the day in work vehicles, often with new items to recycle. For example, several, who appeared to be electricians, would sell new boxes of unopened electrical wire. None of these individuals would engage me in conversation, but came and left very quickly. It was unknown if the wire was stolen or if it was illegally taken from the employer. However, the electricians and others who had similar jobs appeared to function with impunity within the recycling center.

Moreover, observations at the recycling centers revealed many regulars who are employed in fields with access to metal. These individuals never received the same level of suspicion or questions as those who did not have a valid electrical or plumbing license, were not employed by a business with access to metal, or were unknown to recycling center employees. During one discussion with a recycling center manager on the techniques he uses to identify stolen metal, I summarized the tactics of individuals who use work as a cover when recycling stolen material. After explaining the procedure, the manager stopped what he was doing, stared out into the yard, appearing to look at nothing in particular, yet with a facial expression somewhere between alarm and thoughtfulness. Then, slowly turning to me with his eyes widening and his facial expression changing to concern he said, in pensive incredulity, "Those are the guys that I trust!" Apparently, the manager had never considered how individuals with legitimate access to metal may be selling stolen material and was distressed over the idea.

It is hard to know just how often recycling centers are complicit in criminal activity. Ultimately, most thieves said they believed many of the recycling centers knew what they brought in was stolen. They also said there were recycling centers that were known to accept stolen metal without asking questions. In this study, however, with one exception (Eric), the thieves' behavior of concealing their activity from the yards suggested otherwise. Nearly all the thieves attempted to avoid detection at the yard. Clearly, recycling centers play a role in metal theft, as they are the ultimate purchaser of the material. However, with the techniques metal thieves use to mislead recycling centers, it's hard to know to what degree they are culpable.

8.5   SUMMARY

This chapter has identified many of the techniques and methods used for metal theft. In doing so, it has identified three different stages of metal theft: the place, the theft, and the exchange. When considering a place to steal, metal thieves often identified targets based on the quantity and quality (e.g., copper) of metal available. Also considered are places with limited guardianship, such as abandoned property or isolated locations (e.g., farms, forests) along with business with large amounts of metal (e.g., scrap yards, construction sites). These places are selected from the thieves' personal knowledge of a place they commonly visit. When this is not possible, thieves rely on others to assist in determining places to commit theft or may, in rare instances, search out abandoned places online. After a site has been identified, metal thieves tend to case the place in preparation for the theft. This is particularly the case if they must commit burglary, persons are present on the premises, or the place is not known well.

After a place has been identified and cased, metal thieves move to the next stage, the theft. This stage is accomplished in differing ways depending on the characteristics of the place, items to steal, and the thief. Metal thieves are evenly split on committing theft during daylight as opposed to night. Generally, thieves steal metal at night when people may be present and steal during daylight when they needed the light to see or had an adequate explanation for their behavior (e.g., masquerading as a mechanical technician). Thieves tend to steal when they have a financial need; however, a few steal on a regular basis regardless of need. Thieves also tend to return to the same location to steal again if the metal is replaced and the first theft has been easy or in situations where there was too much metal to take the first time. The tools and transportation varied significantly and often depended on the type of items stolen.

Once the metal has been stolen, thieves must then sell it to earn a profit. There are no circumstances when thieves kept the metal; in every case, the metal is sold as scrap for its constituent value. Selling to scrap yards is done in one of three ways; criminal exchanges, surreptitious exchanges, or trusted sales. Criminal exchanges are instances when the scrap yard knowing knew they were purchasing stolen metal. These types of exchanges occurred rarely. The surreptitious exchange takes place when a metal thief takes special effort to conceal the stolen metal from the knowledge of the scrap yard. Surreptitious exchanges are completed by visiting many yards, selling stolen items in various legal jurisdictions, comingling stolen items with legitimately gained scrap, using fences, or by damaging items to reduce suspension by the yard. Trusted sales occur when criminals pose as legitimate workers who would reasonably have access to metals, such as impersonating a HVAC worker or using employment as a HVAC worker who recycles AC units after a job.

REFERENCES

Ashby, M. P., Bowers, K. J., Borrion, H., & Fujiuama, T. (2014). The when and where of an emerging crime type: The example of metal theft from the railway network of Great Britain. *Security Journal*, 1–23.

Bennett, L. (2008). Assets under attack: Metal theft, the built environment and the dark side of the global recycling market. *Environmental Law and Management, 20*, 176–183.

Berg, R., & Hinze, J. (2005). Theft and vandalism on construction sites. *Journal of Construction Engineering and Management, 131*(7), 826–833.

Burnett, J., Kussainov, N., & Hull, E. (2014). *Scrap metal thefts: Is legislation working for states?* Lexington, KY: The Council of State Governments.

Boba, R., & Roberto Santos, L. (2007). Single-family home construction site theft: A crime prevention case study. *International Journal of Construction Education and Research, 3*(3), 217–236.

Clarke, R. V., & Goldstein, H. (2014). Reducing theft at construction sites: Lessons from a problem-oriented project. *Community Oriented Policing Services, US Department of Justice, Criminal Justice Press*.

Cohen, L. E., & Felson, M. (1979). Social change and crime rate trends: A routine activity approach. *American Sociological Review*, 588–608.

Cornish, D. B., & Clarke, R. V. (Eds.). (2014). *The reasoning criminal: Rational choice perspectives on offending*. Piscataway, NJ: Transaction Publishers.

Ecclesiastical Insurance Group. (2014). *Guidance notes: Theft of metal*. Gloucester, GL: Ecclesiastical Insurance Group.

Kooi, B. (2010). *Theft of scrap metal: Problem-oriented guides for police series. Guide No. 58*. US Department of Justice, Office of Community Oriented Policing Services.

Kudla, J. (2012). *Data analytics forecast report: Metal theft claims and questionable claims from January 1, 2009 to December 31, 2011*. Des Plaines, IL: National Insurance Crime Bureau.

McCord, E. S., & Tewksbury, R. (2013). Does the presence of sexually oriented businesses relate to increased levels of crime? An examination using spatial analyses. *Crime & Delinquency, 59*(7), 1108–1125.

National Insurance Crime Bureau. (2014). *NICB: Insured metal theft claims see three-year decline*. Des Plaines, IL: National Insurance Crime Bureau.

Posick, C., Rocque, M., Whiteacre, K., & Mazeika, D. (2012). Examining metal theft in context: An opportunity theory approach. *Justice Research and Policy, 14*(2), 79–102.

Robb, P., Coupe, T., & Ariel, B. (2015). 'Solvability'and detection of metal theft on railway property. *European Journal on Criminal Policy and Research, 21*(4), 463–484.

Sidebottom, A., Ashby, M., & Johnson, S. D. (2014). Copper cable theft revisiting the price–theft hypothesis. *Journal of Research in Crime and Delinquency, 51*(5), 684–700.

Sidebottom, A., Belur, J., Bowers, K., Tompson, L., & Johnson, S. D. (2011). Theft in price-volatile markets: On the relationship between copper price and copper theft. *Journal of Research in Crime and Delinquency, 48*(3), 396–418.

Souza, C. (2015, January). *Ag Alert: Legislators join rural crime detectives to fight metal theft*. Retrieved from http://www.agalert.com/story/?id=806

Stucky, T. D., Ottensmann, J. R., & Payton, S. B. (2012). The effect of foreclosures on crime in Indianapolis, 2003–2008. *Social Science Quarterly, 93*(3), 602–624.

Sutton, M. (1995). Supply by theft: Does the market for second-hand goods play a role in keeping crime figures high? *The British Journal of Criminology*, 400–416.

Washington State Department of Agriculture. (2015, January). Rural Crime Prevention. Retrieved from http://agr.wa.gov/FoodSecurity/docs/238-MetalTheftBrochure.pdf

Whiteacre, K. (2014, November 4). (B. Stickle, Interviewer).

Whiteacre, K., & Howes, R. (2009). *Scrap yards and metal theft insurance claims in 51 U.S. cities.* Indianapolis: University of Indianapolis Community Research Center.

Whiteacre, K., Medler, L., Rhoton, D., & Howes, R. (2008). *Indianapolis metals theft project: Metal thefts database pilot study.* Indianapolis: University of Indianapolis, Community Research Center.

Whiteacre, K., Terheide, D., & Biggs, B. (2014). *Research brief: Metal thefts in Indianapolis October 1, 2011–September 30, 2013.* Indianapolis: University of Indianapolis Community Research Center.

Whiteacre, K., Terheide, D., & Biggs, B. (2015). Metal theft and repeat victimization. *Crime Prevention & Community Safety, 17*(3), 139–155.

Zhang, H., & McCord, E. S. (2014). A spatial analysis of the impact of housing foreclosures on residential burglary. *Applied Geography, 54,* 27–34.

CHAPTER 9

Theft Partnerships and Learning

This chapter focuses on criminal partnerships and how thieves learn the techniques and methods of metal theft. Partnerships between metal thieves are widespread and significant for many thieves. These partnerships enhance a thief's effectiveness and tend to be long-lasting. In addition to discussing the role of criminal organizations, this chapter examines the ways they form and function. The chapter also provides a detailed analysis of how metal thieves learn these methods and techniques. The methods of learning include gaining knowledge from each other, through work, and by trial and error. Each of these learning methods is unique but not exclusive, and many thieves learn by all three methods.

9.1   METAL THEFT PARTNERSHIPS

A majority of thieves interviewed in the present study worked in teams to accomplish thefts. These teams or partnerships ranged from just two individuals to groups of five or more. However, the most common was a group of two persons who worked closely together to steal metal. Often these partnerships were formed outside of, or before, metal theft. For example, Michael and Eric were high school friends; Jessica was romantically involved with her theft partner; John and Matt were each co-workers with their respected co-conspirators; and Dustin collaborated with his brother. These partnerships appeared to be robust and lasting, likely due to the healthy relationship which existed before the metal theft.

© The Author(s) 2017
B.F. Stickle, *Metal Scrappers and Thieves*,
DOI 10.1007/978-3-319-57502-5_9

Moreover, unlike casual relationships with other metal thieves outside the relationship, which were often not positive, partners usually remained loyal to one another.

9.1.1   Going It Alone

While partnerships are common, not everyone in the present study worked as a team. Consider Daniel, who preferred to work alone, was homeless, had no metal related experience, and despite supplementing his scrapping with metal theft struggled to earn enough to eat. When discussing partnerships, Daniel talked about the many times he felt jaded after a partnership when profits were split and he did not receive what he considered a fair portion. Due to this, he explained his preference to work alone, saying, "I don't really trust nobody," and "It's just people who are homeless, you know, they will fight you for a fucking sandwich."

After several interviews with Daniel, I learned that he had been arrested on metal related burglary and theft charges. Daniel had entered a major metropolitan city's police headquarters through a door that was ajar and stolen radios, computer wires, and other small metal items. He then stripped the wires and lightly damaged the other items in an attempt to demonstrate that they had been thrown away. However, his efforts to damage the items were of no avail, as the scrap yard employees suspected that the items, especially the police radio, were not found discarded in a dumpster as Daniel had claimed. Daniel was quickly apprehended after a police officer recognized him from video surveillance footage. While visiting Daniel in jail, shortly after his arrest, he explained the burglary:

> *Daniel:* I went down the police headquarters and got some scrap parts, and I scrapped it. I don't know what the hell I was thinking.
> *Interviewer:* What did you take?
> *Daniel:* Just some computer parts and stuff and [I]scrapped it. That ain't right, I know. Now I am looking at burglary three!
> *Interviewer:* So what prompted you to do that?
> *Daniel:* I don't know, man, I got tired of living on the streets.

Metal Thief Profile

Name:	Daniel
Taxonomy:	Metal Thief
Location:	Southeast—Inner City
Age:	32
Race:	White
Education:	Some College
Employment:	Unemployed
Experience:	I met Daniel several times and talked with him often. He walked the inner city streets and picked up any metal he could find and only worked alone after he'd had issues with scrapping partners. The day I first met him he appeared to be high, going by his severely slurred speech, incoherent babbling, and the snot and saliva constantly dribbling from his nose and mouth. As I asked him questions he would become fixated on a topic and repeat the same story, making it difficult to have a conversation. He was dirty, smelled very bad, appeared emaciated, and his teeth appeared to be rotting in his mouth. He also showed me open wounds around his ankles where it appeared that something was eating away at his flesh. While I was observing him, he looked intently at a mangled metal chair and said to me, "Yes, sir. Please, 'cause I just need help learning," indicating that he wanted me to help him learn how to take the metal apart and sell it according to type for a higher price. I was happy to help and he was thankful for my assistance. After he left the yard, one of the staff came running out looking for me. She sighed when she spotted me and said, "Wow, we thought you left with him to talk, don't do that with that one, he's dangerous."

I talked with Daniel quite often during the course of the study and all the other times he appeared more sober; however did appear to have a mental illness, which at one point he addressed by saying, "They used to think I was schizophrenic and bipolar (long pause), which I wasn't." However, a few days later he described how in the past the jail put him on lithium, which is a common treatment for schizophrenia and has been known to damage teeth.

Just as with Chris (see Chap. 7), I had tremendous pity for Daniel, and since he traded his food stamp card for rent, I occasionally bought him food. I got a phone call one day from the yard that he frequented and they told me he had been arrested for metal theft. Apparently, he had walked into the police headquarters, stolen various metal items, and scrapped them. My last interview with him was in the jail where he appeared, for the first time I knew him, in his right mind. He had gained weight (nearly 20 pounds) and looked much better. I asked how they were treating him and he said, "Well, it's better than sleeping in the streets and it's dangerous on the streets. And at least I am eating three meals a day … I'm on the psych ward. They found out I was bipolar!"

Daniel was not alone in his unsystematic and opportunistic approach to metal theft, or in his desire to work individually because of trust issues related to profit sharing. Chris, who is also homeless and struggling to eat despite his metal theft income, also expressed frustration when he tried to collaborate with others, saying:

'Cause, everybody wants something for something. Somebody helps me do this, and I don't do it by myself, they are like, oh, I helped you; I helped pull this out, and I helped do this, where is my half at?

Working without a partner is usually by choice, according to the thieves interviewed. Moreover, those who functioned individually seemed to commit more opportunistic metal thefts (e.g., Daniel and Chris), to be unskilled (e.g., no previous employment as a plumber or electrician), were generally less successful thieves, were more often questioned and arrested by the police, and distrusted others.

9.1.2   Through Thick and Thin

Two of the primary concerns among those who prefer to work individually related to disputes in profit sharing and a lack of trust; those who worked in teams rarely reported having such issues. Moreover, those who functioned in partnerships tended to maintain their affiliations for extended periods without serious disputes. In fact, in the present study, conflicts between established partners seemed rare, even over money. Perhaps this is due to the extensive relationships that tended to exist before stealing, because teams appeared to be more effective and the profits greater, or due to the usual processes of evenly splitting profits. Regardless of the reasons, disputes were rare. Nearly all teams split the profits evenly unless there was a theft-related expenditure by one of the members (e.g., one member provided the vehicle). Other teams functioned so efficiently that the partnership resembled a business operation, paying expenses related to the theft before splitting profits, and having arrangements for how the proceeds were split based on who knew the place to find metal. For example, Leo described his relationship with his partner thus:

> *Leo*: I always [stole metal] with me and another guy, you know, so it was just the two of us. If we made $5000, we would [each get] $2500 apiece, after we put more gas in the truck and [bought] more duct tape and stuff like that.
> *Interviewer*: Okay, so you all treated this as a business, right?
> *Leo*: Yeah, pretty much, we made sure the truck had gas, if there was anything wrong with the truck we would make sure to get it fixed and make sure we buy new hacksaw blades, new tools, you know what I mean, that all comes off the top, and then we split the rest.

When disputes did occur within partnerships, they often resulted from the arrest of one of the partners. These incidents seemed to stress the relationships and caused fissures, a few of which were permanent while others were temporary. For example, Dustin shared an incident when he and his partner (his sibling) were charged with metal theft. While they were stealing an air conditioner, the police unexpectedly arrived. Dustin was able to flee without apprehension, but the police captured his brother after a short foot chase. After capture, Dustin's brother was interrogated by police. To earn enough money to bail his brother out of jail the following day, Dustin sold stolen metal from a previous job. Once they were home, Dustin learned that his brother had told the police of his (Dustin's) involvement in the theft:

> My own brother snitched on me, and that hurt me the most! He said he did it because the cops were stomping on his head. It should not have mattered, but he did what he did. I got so mad he seen the anger in me, he turned so white, I started to go after him (physically) and my wife and his wife stopped me and he ended up urinating on himself, but we got through that, and things just went the same. And yes, I did dirt with him again. It's just a brother's love, I guess.

The partnerships established before involvement in theft or on a more intimate level, such as siblings or longtime friends, were more likely to be resilient than connections formed in other ways. For instance, John and Michael established a criminal enterprise with over a dozen people on their work crew. When the metal theft was discovered and investigated by law enforcement, John and Michael both blamed their other partners for talking to law enforcement, which lead to their arrest. While Michael and John remained friends, likely due to a relationship established before the theft, none of the other relationships recovered. Actually, Michael was so angry with one of his partners he received an additional charge because he was found to have threatened his former partner. As he recounted it:

> My boss and that so-called "friend" did no prison time. My friend was the first one to go in and admit to everything. So when I found out about it, I wrote on Facebook that he would "get his in the end," meaning karma. Guess what? The next day six [law enforcement officers] arrested me for intimidation of a government witness. They gave me an extra year because of that.

Notwithstanding the limited instances of partners turning on each other during law enforcement investigations, partners usually remained loyal to one another. This loyalty and positive relationship appeared to endure throughout the course of the criminal partnership and even beyond.

9.1.3   Criminal Enterprise

Interestingly, the partnerships in the present study did not seem to have an internal hierarchy. While this topic was not specially addressed in the questioning, none of the thieves who had collaborates expressed or indicated that there was one or the other who led the team or was in charge. In fact, there were often stories suggesting leadership roles would change depending on availability of vehicles and, more important, knowledge of places with existing metal. The leadership role would often transfer to the thief who had the best knowledge of the place to be stolen from or had the knowledge necessary to commit the theft.

For example, Jessica began to be involved in metal theft by simply allowing her boyfriend and his family and friends who stole metal to use her truck, saying, "I never did go, but I let people use my truck to do it. So I would still get money out of it, because I let them use my truck. That is mainly how I got involved in it." As time went on Jessica began to be the driver. She said, "Either I would sit in the truck and wait on them, or you know—" with her voice trailing off and eyes widening, indicating she began to go with them and participate in the thefts. Jessica primarily served the role of being the one in the group of thieves who "had the license, insurance, and truck" and therefore was not usually the leader, but still received an equal share of the money when "it was divided up between how many were there." In time, however, Jessica began to suggest locations where she knew of valuable metal, either hearing about it from an acquaintance or through work. In these instances, she took on the lead role of organizing the theft. For example, Jessica had past work experience at a local vehicle junkyard and one evening lead her group when they trespassed and stole catalytic converters from junked vehicle. Due to her previous employment on site, she provided her group with knowledge of when to steal, the location of the catalytic converters, and how to defeat the security measures (cameras, lights, fences, and the like). After the theft at the junkyard, she resumed her primary role of driver for the group.

The way that Jessica meandered within the group, providing different roles dependent on what was needed or the experience and knowledge she possessed, seemed common to many of the other groups. Regardless of how the group formed, the leadership roles were always directed toward the individuals with the most experience with the item to be stolen (e.g., air conditioners) or the person with the best knowledge of the place to steal from (e.g., Jessica and the junkyard).

In addition to the changing leadership roles of the criminal enterprise, the individual tasks completed by each partner also changed. As identified earlier when discussing leadership, tasks such as driving fell to whoever in the group had a license and a reliable vehicle. For example, David explained they would select a driver from the group who was not known to the police and had a functioning car, describing how the police would "usually get you for your license plate [having] expired, a [head] light out, or the cop recognizes you or knows you are wanted or something like that." Thus, the job of driving the car would fall to the person who had the vehicle least likely to be stopped by the cops.

Similarly, other tasks were delegated according to the experience the thief had. For example, Leo explained that when he had a risky job or was teaching someone who was new to metal theft, he would assign him or her as the lookout while the other experienced thieves stole the metal, saying, "So you just had someone else pretty much looking out the window while you're doing what you're doing. You'll have a radio, he'll radio you, and he'll tell you to stop making noise when he would see the patrol car coming around." Other tasks for new thieves might include hauling the metal from inside a building out to the vehicle, stripping the rubber coating off the copper wires, or transporting the metal to the scrap yard.

Some of the tasks within the group were not based on experience at all. For example, Michael's theft partnership included the crew he worked with as an electrician. Michael explained how he brought his friend, Eric, into the group because he and his work crew were stealing so much at work that he could not keep up with stripping and taking it all to the recycling center:

> In the beginning, it was only crew members, but then the last two years it was my best friend [Eric]. He just happened to be at my house one day when my boss and I were stripping some [cable], and he asked what we were doing. We told him, and he asked if he could help, and we were cool with it. Then it was easier and more practical for him to take it in every week. We would just split it [the profit].

Eric explained his role of splitter, transport, and seller of the stolen metal from his perspective:

> *Eric*: My involvement was basically I would ride over with those guys (on Michaels work team), it would be Michael and all of the guys from [the company], we would turn it into a recycling center, and basically they got familiar with me. My whole deal was to show my face, let me be seen with these guys because the recycling center was used to seeing those guys.

So once they started seeing me with those guys, they kinda started con-
necting me with them and then because Michael and the other guy that
was driving wanted to kind of stop showing up at the recycling center
just to be on the safe side. So basically they had asked me to drop it off.
So that's more or less [it], Michael would drop off what he [and his work
crew] got that day in my garage, and then I stripped it, and I loaded it
up, and I took it to the recycling center.

Interviewer: Okay, how did you cut the money, if you don't mind me asking?

Eric: Between us, we split it, because he (Michael) got it (the metal), but I
did a lot of the work. We justified, well, I'm stripping it, that's a ton of
work, and I'm hauling, so we split 50/50.

Interviewer: How long would it take you strip the wire in an average week?

Eric: It could be three or four hours a week, actually it was a lot. I mean I'd
have to be out there every single day just to make sure I got it all.

In Michael and Eric's case, Eric was brought into the partnership
after the theft had already occurred and his job was simply to strip
and sell the wire. Eric had no involvement with the actual theft of the
item. Eric was also used to shield the group that was responsible for the
actual theft, as Eric described above by explaining how he mingled with
the group that was stealing until he was known at the recycling center.
After that the group did not return there, and Eric was the one making
the sale, thereby reducing the suspicion on the main group. The transfer of suspicion or level of criminal activity being shared among a metal
theft group is not unusual either. For example, Zach and his metal theft
partner decided that Zach would be the one to enter a property to steal
the metal, then throw it to his partner in the truck outside the fence.
This decision was made because Zach's partner had previously been
arrested on burglary charges and would be punished more severely if
caught.

The task assigned to each person in the group appeared to change
based on a variety of circumstances. Leadership roles were primarily based
on knowledge of the place where the theft occurred or having a specialized skill to accomplish the theft. Moreover, the tasks assigned to each
member of the group would depend on the resources they possessed (e.g.,
a vehicle), if they were new or an unskilled at metal theft, or if they had
a pending or recent charge that placed them in greater danger of punishment if captured. Finally, a few groups maintained a firm structure of who
did what and had individuals who took part in only one aspect of the theft,

such as Eric, but this was rare, as partnerships tended to trade leadership roles and tasks.

9.2   How Thieves Learn

Since most thieves tend to drift into metal theft (see Chap. 7), it is important to study how they learn the necessary techniques and methods. Based on the present study's findings, thieves learn to steal through three methods: from each other, while at work, or through trial and error. These methods of learning are not exclusive to one another; many metal thieves learned by more than one method during their thefts. For example, one thief may have initially learned how to steal metal from a friend who stole, but later discussed how he learned to steal certain metals from specific places by trial and error. For ease of comprehension, however, these learning practices are presented here as separate methods.

9.2.1   From Each Other

The overwhelming majority of metal thieves learned to steal from another thief. Learning from another thief should not come as a surprise, as the present study has demonstrated that the tendency to work in partnership with one another in metal theft is very common. Typically, relationships form before criminal activity. Then at some point, usually due to financial need, the relationships transformed into a metal theft partnership. For example, when Chris and his girlfriend exhausted their finances, his girlfriend suggested he steal metal with her brother telling him, "You can go scrapping copper and stuff like that." Chris explained that his girlfriend's brother stole metal "all the time," and taught him how to steal.

Similarly, Jessica described how she began dating a man who stole metal with a group of other individuals. Her first participation was as the driver. She explained, "That is mainly how I got involved … I would sit in the truck and wait on them." While Jessica was the only licensed driver with a vehicle in the group, over time she progressed from being the driver to learning more about metal theft by "just being around the guys" and began to engage in burglary and trespassing to commit metal theft on a regular basis.

Jessica spent years stealing metal with her boyfriend and "the guys," learning a significant number of techniques and methods. Due to these experiences, she learned how to be a successful thief and was able to produce "$70 and some days we might make $300" with little time commitment. However, Chris' relationships with his girlfriend, and by extension her brother, were short-lived and he did not learn as much as other thieves did. For example, despite squatting in an abandoned house and stripping its metal contents, Chris struggled to "put $10 in my pocket today."

Based on the information gathered in the present study, the best method for learning to be a successful metal thief appears to involve extensive time with other, experienced thieves. A quick theft here and there or brief instruction by another thief is typically not sufficient to teach thieves how to be an efficient metal thief. On the contrary, a longer relationship often centered on a partnership with another skilled metal thief is usually necessary for a new metal thief to proficiently learn the techniques and methods and be successful at metal theft.

Some of the thieves interviewed also discussed teaching other thieves the approaches and techniques of metal theft. For example, when Michael became concerned that his repeated appearance at recycling centers would tip off law enforcement, he brought in Eric as a partner and taught him how to strip wires. Similarly, Leo was faced with a difficult decision when his longtime partner was arrested and sent to prison. Rather than stop stealing, Leo taught a new partner, sharing the story this way:

> I had my best friend. We were doing it for a while and then he ended up going to prison for unrelated issues. I ended up meeting up with this other guy that I became friends with, and I showed him how to do it, you know, and we were doing it, and then when he got out of prison we ended up picking back up, per se.

Even when thieves began stealing without a partnership, their skills were advanced when they worked together with another thief and learned new techniques and methods.

9.2.2   At Work

In the present study, the most successful metal thieves also had training and experience gained from legitimate work, often with metal. As described previously, nearly all metal thieves were currently employed, or had previously been employed, in a field related to working with metal (e.g., construction, employment at a recycling center). Employment provided the necessary background and expertise needed to steal metal successfully. Specifically, it allowed metal thieves the opportunity to gain technical skills, increase knowledge of particular types of metals, and often provided the tools necessary to carry out large and technically challenging amounts of metal theft. Additionally, regular employment in a metal related industry often exposed metal thieves to places with high quantities of valuable metals and provided a ruse to anyone suspicious of their actions. However, it is important to distinguish between learning how to steal and learning about metals, both of which enable individuals to steal. Both types of learning occur among metal thieves.

Some metal thieves experienced circumstances at work where they were taught to steal by a co-worker. For instance, Matt's co-worker at the HVAC company they worked for showed him how to steal an air conditioner quickly. Matt's co-worker even took him along on several thefts to demonstrate and further teach the techniques. Matt described how he learned to steal:

> *Matt:* I can remember the first time I did it. I was scared to death, but I needed the money, I needed the money bad. And, my partner at the time, [an] older gentleman, he had, I mean it was nothing, and he had done it for years. He said, "The same way we put them down is the same way we take them out! Man, you go back there, we are going to pop the line set, you are going to cut the box, there's two wires and if there is a disconnect box you gonna pull the disconnect, cut the hot, yank it out, cut the line, if it ain't bolted down, we out of there."
>
> *Interviewer:* So he helped teach you.
>
> *Matt:* He helped me get the courage, 'cause I was like— (demonstrates a facial expression of concern while shaking his head in negation).

In Matt's case, he was taught to steal an air conditioner unit by reversing the order in which he installed it. By working backward, he was able to quickly disconnect an air conditioner and was able to steal several within a short time. The skills he learned installing them for customers were the same skills he used to steal them after his shift. Moreover, much of what he learned from his co-worker was the application of the skills he already possessed and, primarily, the encouragement to commit the crime. In fact, learning from another experienced metal thief by watching or participating in an initial experience of theft was commonly discussed.

Metal Thief Profile

Name: Matt
Taxonomy: Metal Thief
Location: Southeastern—Urban City
Age: Mid-30s
Race: Black
Education: Bachelors
Employment: Full-time—HVAC Installer
Experience: One of the first things I did as I began to conduct this research was send a mass email to all my friends and acquaintances, describing the project and asking if they knew anyone who scrapped or stole. I was surprised when a friend of mine said, "I need to talk with you about that." We set up a meeting and I was surprised to learn of his background. While I knew Matt had been active in the drug scene in the late 1990s, I had no idea it was largely funded by metal theft. Matt described how he was working for his father's HVAC company installing new air conditioners when his partner at work taught him how to steal the units and recycle them for money. He described how he would use his work as a cover, both in case someone saw him and with the yards, so that they were not suspicious of him. He would use his time installing an air conditioner unit to scout out nearby places with large units (worth more money) and would come back after work. Although he was never caught, he believes he stole nearly 100 units before he was arrested on unrelated charges. While in jail awaiting trial, Matt explained how he had a spiritual experience and that when he was out he changed his life around. Matt now has a successful career, is married, a minister, and recently graduated with a master's degree and has started on his PhD.

Comparably, Michael explained how he was taught to steal electrical cable by his co-workers and boss on the first day of work:

> From day one, we were told that [recycling cable] was one of the benefits of working [at the company]. At the end of the job, our lead man, who would later become a boss, said, "We're gonna take it [the cable they replaced] to another co-worker's house and burn the lead off of it,'" and then he would cash everything in and split it between the seven of us. We each got around $700 for it! I was like, wow, this is awesome!

However, not everyone was taught how to steal while at work. Work also helped metal thieves learn to steal by educating them on the value of metal and the process of how to recycle it. Learning from employment experience which metals had more value and how to handle them was commonly discussed. This knowledge was then used to identify and process stolen metal. For example, James, who often trespassed to steal metal, explained how he learned to recycle by saying, "Actually ... I worked roofing, and when we [tore] off some of the metal from the roof we knew it was copper or aluminum, so I learned it [that way], you know what I mean?" Similarly, Zach initially learned about metals when he worked for his father's automobile junkyard and later used that knowledge to steal from businesses that had large amounts of metal stored outdoors.

Metal Thief Profile

Name:	James
Taxonomy:	Metal Thief
Location:	Southeast—Urban City
Age:	Mid-50s
Race:	White
Education:	Unknown
Employment:	Unemployed
Experience:	I met James inside an inner city dumpster on a cold winter's day. We began to talk about scrapping and he described himself as a professional scrapper. He had an old dilapidated truck parked nearby full of metal parts, and his girlfriend sat inside while he searched

for metal. He was a rough-looking man who was far more interested in looking for metal than talking to me. James made most of his income from scrapping, saying, "I can get on down the road and before the day is up I can pick up, I'll have almost a ton and I'll make 200 dollars today by five o'clock." While James claimed to not be a thief, I had my doubts based on his described activities. He also had difficulty explaining how he knew what was set out for trash and what was being kept by the owners. James described several times when he would jump a fence to get metal in the back of a business that he claimed the metal was "going to the dumpster anyway, [they] just don't want us to have it." Unfortunately, the more we talked, the more agitated he seemed to become with my questions, and abruptly left.

While a few individuals learned to steal metal from co-workers, the majority utilized the skills, knowledge, or tools gained at work to locate, steal, and exchange the metals. In each case, having been employed in a field related to, or exposing an individual to, metal enhanced the metal thieves' awareness of the metal's value and knowledge of the methods necessary to obtain it. Moreover, those employed in the field (actively or in the past) tended to yield higher quantities and qualities of metal when stealing.

9.2.3   Trial and Error

Trial and error were also commonly spoken of among metal thieves. Some learned exclusively through this method and had little work experience or practical guides to teach them. Leo explained that he had legally scrapped a few times, but had never stolen metal until he learned about it from news coverage:

> *Leo:* I actually kind of learned it on my own. I was watching the news one day [and] this guy got arrested. He stole millions, and millions of dollars; it was ridiculous, for stealing copper out of an abandoned building. I said

Image 9.1   An improvised ladder to enter the large dumpster where I met James

well I'm going to look into this a little further and I did, you know, and I ended up just starting to do it. The first time I did, it was scary, but you know I wasn't really good at it, but over the years I got really, really good at it.

Interviewer: How did you learn?

Leo: Actually, that was trial and error. That type of thing. The first time I did it I just kind of brought everything to the scrap yard and the scrap yard was like, "Oh, you know you can get more money if you do this," and I was like, "Oh, really," so now that's why we start cleaning it and stuff.

Leo went on to explain how he learned different techniques and methods for effectively stealing metal by experimenting. He was not alone in this regard. Many of the thieves interviewed explained or described how they continued to learn or refine techniques for stealing metal through trial and error. Dustin spoke of how he and his brother learned to scrap and eventually to steal more efficiently, saying, "We just learned from our mistakes, and we learned real quick what was worth more money, so we look[ed] for them items first, but whatever was worth money we took."

Another method of learning by trial and error occurs after the theft but before exchanging the metal for money. Many thieves shared experiences of learning how to separate metals more efficiently to get a higher price. For example, stripping the plastic coating from electrical wires or removing soldered connections off copper pipes can double their value because different types of metal are not mixed. Sometimes this knowledge occurred exclusively by trial and error, such as when thieves received a lower price for a bucket of electrical wire with the coating on it, compared to a bucket of wire that had been stripped of the coating.

In other instances, a recycling center employee would make suggestions or take the time to teach them how to process their metals so they could earn higher profit efficiently. Regardless of the methods by which thieves learned to process metals better, the result was a higher price and thus encouragement to continue stealing more. For example, Daniel, who was attempting to clean a load of suspiciously acquired metals at the recycling center, explained:

> They [scrap yard employees] had to learn me; for real, they taught me a lot. Yeah, they taught me a lot, like to break the shit down all the way. When I wasn't making no money, you know how you get frustrated being out here all day and not making no money. (Turning to an employee) Is that aluminum? Sir, please … I just need help learning!

The instruction that Daniel received from the recycling center employees each time he arrived allowed him to earn more money from the metal he collected and stole, as well as encouraged him to continue collecting metal.

However, it should not be misconstrued that recycling center employees know they are assisting metal thieves who are learning to be more efficient. Rather, I observed recycling center employees frequently offering advice and teaching customers how they could increase their profits

Image 9.2   A yard employee helping to unload heavy metal items

by processing the metals better. For example, an employee might explain while weighing copper wire that if the customer removed the plastic coating (stripping the wire), they would receive a better price for copper that is "clean." This communication and education appeared to be standard practice at most recycling centers and was a method thieves learned to grow their profits.

9.3   SUMMARY

This chapter provides several insights into the partnerships metal thieves form and how they learn to steal. A few metal thieves operated alone. Usually, according to the thief, working alone was a choice and primarily revolved around trust issues with other thieves. Metal thieves who worked alone tended to commit more opportunistic metal thefts, to be unskilled, were generally less effective thieves, were more often questioned and arrested by the police, and distrusted others.

Most metal thieves operated in groups, and many maintained a criminal enterprise of thieves who worked together. This partnership was long-lasting and robust, commonly based on a relationship outside their criminal activity such as romantic relationships, family, co-workers, or close friends. Thieves in partnerships tend to be more efficient, stole in greater quantities, and were more calculated in their crimes. The partnerships had little formal structure and contained at least one individual who had experience in a related metal industry. Leadership roles within the group alternated according to who had knowledge of a place with metal and more experience in the technique used to accomplish the theft. Moreover, the tasks or roles within the group also changed according to who was new to the group, who had a functioning car, and other factors that determined what each individual did during the theft. Partnerships were long lasting and stable. However, if one partner implicated the other when interrogated by law enforcement, the future of the alliance was in jeopardy.

Thieves employed a wide variety of learning techniques, including learning from each other, through work, and by trial and error. For example, some thieves were exposed to the knowledge of a metal's value while at work, but did not learn the techniques unique to theft until they developed a friendship with a metal thief. Other thieves learned through trial and error. Some thieves were brought into a partnership and taught the methods, but continued to learn through trial and error. Still others were taught how to steal by co-workers. Regardless of how thieves initially began stealing, each of these learning processes was often present, to one degree or another, and continued for some time. In addition to learning to steal, several metal thieves also discussed actively teaching others to commit metal theft. Consequently, the cycle of thieves teaching others how to steal, combined with continued learning by way of trial and error, means that thieves continue to create more metal thieves who will continue to possess better-honed methods and techniques, which enhance profits and lure them into more theft.

CHAPTER 10

Social Controls of Metal Thieves

The previous chapters examined the methods and techniques by which thieves commit theft, as well as how they learn those techniques. This chapter identifies social controls that may influence metal thieves and examines how metal thieves respond to these social controls. According to F. Evan Nye (1958), social control describes an internal means of control, such as values, norms, relationships, moral codes, beliefs, and commitments that encourage individuals not to violate the law. Nye proposed that social control is expressed in three primary areas: direct social control, indirect social control, and internal social control. This chapter will address each of Nye's proposed areas of social control, providing an understanding of how social controls are experienced by a thief and how they influence a thief's behavior.

10.1   DIRECT CONTROL

Direct control is the process by which punishments or rewards are given to an individual based on their behavior. Typically, this occurs when individuals deviate from accepted norms and are punished (e.g., receive jail time for a theft), or when people meet society's expected norms and are rewarded (e.g., receiving a bonus at work for excellent performance). However, since metal theft is a deviation from societal norms, most metal thieves are concerned about the punishment they may receive when caught. For example, most metal thieves stated that they did not tell other family or friends about their metal theft behavior; the majority of direct control, therefore,

© The Author(s) 2017   193
B.F. Stickle, *Metal Scrappers and Thieves*,
DOI 10.1007/978-3-319-57502-5_10

comes from the fear of being investigated, arrested, or punished by law enforcement. This fear was evident in several ways, including concern for dogs, cameras, witnesses to the crime, and discovery by law enforcement.

10.1.1   Dogs

Most metal thieves discussed dogs and the deterrence effect they had on their behavior. For instance, when Dustin, who stole hundreds of air conditioners from occupied homes, was asked how dogs affected his behavior, he whimsically stated that they are "too loud. I was crazy, but not stupid!" Likewise, while I was walking in a neighborhood with Chris, who was pointing out to me the attractors and detractors of different houses in relation to metal theft, he spotted a dog, and this brief exchange followed:

> *Chris:* Look at that Doberman!
> *Interviewer:* What about that, would that keep you away?
> *Chris:* Yeah! Hell yeah! That's a big dog right there!

While Chris did see a large and potentially aggressive dog, interestingly, the size of the dog did not appear to be a factor for most thieves. Rather, they seemed to be more concerned with the attention the bark would draw than with the bite. For instance, David explained why he avoided dogs, saying, "Oh yeah, 'cause that would've [woken] somebody up" and indicated that dogs are "definitely a deterrent." Matt also expressed a disdain for dogs, identifying them as one of the things that would "keep me from going in." In fact, the presence of a dog, more than cameras and even nearby witnesses, is a nearly universal deciding factor against theft.

As a matter of fact, when I asked Jessica for advice on keeping abandoned homes from getting broken into and the copper stolen, she replied, "They would need some kind of security or dogs. 'Cause boarding it up, locking it up, that don't work. People be in there anyway." The consensus among thieves was that dogs, either at the theft site (e.g., junkyard dog) or at a nearby location (e.g., neighbor's house), were reasons not to steal because they might alert someone to the thief's presence.

10.1.2   Cameras

In addition to dogs, several metal thieves discussed cameras; however, cameras had a rather mixed effect on thieves. Nearly half of the metal thieves mentioned cameras as a factor that influenced their decision to steal or not steal metal. Those who said it was a factor looked for cameras while casing the place, while others stated that they would steal even if cameras were present. For example, Dustin explained that he "didn't care about cameras [because we] wore face masks." Similarly, Matt explained how cameras would have made him second-guess his actions, but "eventually [I] would have snatched [the air conditioner]." In contrast, a few metal thieves said something similar to Leo, who stated, "Yes, definitely," cameras would prevent him from stealing. As with dogs, the controlling effect of cameras was the potential to be identified after the crime. Several thieves discussed wearing a mask or operating at night to avoid a camera identifying who they were.

In fact, a few thieves mentioned poorly functioning cameras, especially during evening and dark hours. Jessica shared a story where she tipped off a metal theft partner about several vehicles in the place where she worked and suggested he steal the catalytic converters:

> We had eight vans [at work]; well, this guy I knew he went and stole the catalytic converters off all eight vans one night. He did not realize he was on camera. But the camera caught him, but they could not see his face, they could not tell who it was. So he never did get charged with it. You see what I am saying?

Jessica became aware that the cameras captured her friend when she returned the next day to work and saw management reviewing the footage. However, incidents like that seemed to embolden her future actions and her decision to not take cameras seriously.

Another attribute of the controlling effect of cameras was whether the cameras were visible. Several metal thieves noted that if there were a sign indicating that a camera was in operation, they would look for a camera. If they were unable to identify a camera, they assumed the sign was false and would often continue with the theft. Jessica disclosed the following information:

> *Interviewer*: If you saw a sign that says "Cameras in Use," would that have kept you from stealing?"
> *Jessica*: No, we would have just hung out looking to see if we could spot them.
> *Interviewer*: You would just hang out, and if you couldn't see them you would just assume it was not there and say, oh well?
> *Jessica*: Yep!

In the case of metal theft, it appears that cameras are less efficient social controllers than dogs.

To be effective, cameras needed to be seen by the thief and of sufficiently good quality to identify the thief. However, even in situations where the cameras are observed, some thieves take precautions (e.g., masks) to hide their identity, but are not deterred. As David explained, "Some of these guys, it just don't matter. The guys that have a little common sense (cameras) may (effect their behavior), but it really wouldn't matter." In summary, visible cameras seemed to cause thieves to second-guess their actions or take precautions to avoid identification, but in the end did not deter them from criminal activity.

10.1.3 Witnesses

While many metal thieves discussed dogs and cameras, the most common direct control was a concern of witnesses. Regardless of who might see the criminal act, metal thieves were very concerned about other people observing their actions and notifying law enforcement. In particular, irrespective of the place and type of metal stolen, metal thieves universally mentioned neighbors as a risk factor and thus a direct control over their actions. When casing a place, Jessica looked for "nosy neighbors," and many other metal thieves described active or observant neighbors as a factor that would often cause them to alter or abandon their plans of theft altogether. For example, Matt explained how he evaluated the presence of neighbors, and how busy and observant they were; even their gender affected how and when he stole air conditioners:

> We would scope out the house, we knew when [the neighbors] was gone, when they were home, if they were [often] outside or in the house, single mother next door just with some kids or something. If no man's around, you know—(indicating he would commit the theft).

Even if there were neighbors or occupants, a few metal thieves would continue with the theft. However, they would take extra precautions to avoid discovery. For instance, Dustin, who said that nearly 90% of the instances when he stole metal was from occupied homes, described how even if an air conditioner were "just two feet from the window (of an occupied residence he would) still steal it regardless." However, he explained that air conditioner units at occupied homes have to be, "In the

Image 10.1 An air conditioner broken down and waiting to be weighed

open (not surrounded by a fence), [yet] concealed somehow, by bushes or behind the building." Moreover, he also described how the presence of a potential witness or victim would cause him to alter his plans, stating he would return "at night ... when they were asleep." Therefore, the few thieves who did steal when a potential witness was present had criteria for how and when they would commit the crime (e.g., at night, behind bushes, with easy access) to reduce the likelihood of being observed. In most instances, if these conditions were not met they would abandon the attempt and look for a more suitable place.

Other thieves exclusively sought out buildings or areas with no one around, primarily to avoid witnesses. Chris, for example, explained how he selected abandoned homes to burglarize and steal metal with no neighbors around:

> *Interviewer:* And you said you don't want neighbors around you?
> *Chris:* Right, right, yeah, you don't want neighbors, if you're going to go in one, if people are living on both sides of you, they hear you in there.
> *Interviewer (pointing to a home that is empty with an occupied home next door):* So this one is vacant and you will be less likely [to go there] because there is a neighbor?
> *Chris:* Yep!

Image 10.2 An abandoned building with no nearby neighbors makes a prime target for metal thieves. Here the owners have even spray painted "no copper" in an attempt to dissuade thieves from entering

While metal thieves were cautious to avoid discovery, many thieves often had a story or excuse ready for neighbors or others who might observe and question their activities. The most frequently developed story revolved around presenting himself or herself as maintenance, construction, or HVAC installer completing a job. For example, Matt said, "We could pose as a Heat and Air person like we back there fixing the air." Similarly, Leo had construction company logos on his truck to reduce suspicion.

10.1.4   Law Enforcement

While thieves were concerned with dogs, cameras, and other people who may observe their activities, they were most concerned about law enforcement and specifically the possible penalty of arrest. Approximately half of those in the present study were arrested for a metal theft related crime at some point. In most circumstances, the arrest followed an investigation that began with the capture and interrogation of a partner and culminated when a detective showed up at their home. For instance, Eric, Michael, and John described how police "showed up at my door" while investigating a metal theft incident. In a few other instances, thieves were captured while stealing metal. For example, Zach was caught stealing from a sign shop when police responded to a motion alarm. Leo was caught when his truck had a flat tire and a police officer stopping to help observed more than "$6000 worth of copper in the back of my truck."

While some metal thieves are arrested, not all the contacts with law enforcement resulted in an arrest or even suspicion. Some contact with law enforcement had little to do with metal theft and was merely routine (e.g., traffic stops), during which their illegal activities were either not uncovered or were explained away. For example, Dustin described when he was stopped by police while driving late at night with a stolen air conditioner in the back of his vehicle:

> I got stopped with [an] A/C unit. I was coming home from [stealing] it, and I told them I took it from work, [because] it was bad. So, they called a number I gave them, and a maintenance guy told them, yes, it was OK. The [maintenance man] was my father-in-law; he did work for an apartment complex, but not the one I [stole] it from. I was stopped 'cause it was late. That's all, routine … [the cop was] just doing his job.

Other times law enforcement did not notice the crime. For example, Leo mentioned there were several occasions police officers pulled into the parking lot of a building he was stripping, but never saw him. He described one situation where his truck was on site, and he had finished stripping "about three or four thousand dollars' worth of copper" when a police officer pulled into the lot of the abandoned building:

> [The cop] went in the back [of the abandoned building I was in], he walked his dog, and he was just sitting there in the parking lot watching for speeders. Had nothing to do with us, but we got so paranoid at that point, we just kind of hid in the building for a little while… So, I'm in this building smoking cigarettes like crazy, looking out the window of the top floor of this building, staring at this cop for hours in December, freezing my butt off, waiting for him to leave.

Metal thieves were always concerned about contacts with law enforcement, and many took great care to avoid detection. These actions involved concealing items while transporting them, damaging things to lessen their obvious value, and developing a cover story for possession of metals. For example, Michael had an explanation ready in the event law enforcement questioned him, since he had a ton of stolen industrial-sized copper wire:

> We always had stories made up just in case. We had one guy's business card that actually gutted out old buildings, and he told us if we [were questioned by] the police [we could] call and he would cover for us.

For most metal thieves direct control is centered on law enforcement and the potential for arrest. To avoid this punishment, many thieves take the precautions previously described. These often involve avoiding dogs in nearly all situations and typically avoiding cameras when they are visible. In virtually all circumstances, other people were avoided by seeking abandoned properties or operating a night when the occupants were sleeping. Moreover, if confronted, many thieves had stories or justifications prepared to escape suspicion.

Despite frequent contacts with law enforcement, only half of the metal thieves experienced an arrest for metal theft, and a few spent time incarcerated for unrelated charges. All metal thieves took precautions to avoid observation and obfuscate investigations; however, the potential of detection and punishment was not a sufficient direct control to alter long-term criminal behavior. Most metal thieves who had been arrested and convicted for a prison term of over a year or more described the experience as a factor that significantly influenced their decision to steal in the future. In fact, the most prolific and profitable metal thieves interviewed, such as Jessica, Matt, Leo, and Michel—whose combined theft of stolen mental during their active years would probably be valued at several million dollars—described how, after an arrest and subsequent imprisonment, they had not returned to metal theft even after being released. In other words, the constant fear of being discovered and the effects of arrest did not seem to have a long-term controlling effect on the individuals unless a lengthy prison sentence was involved.

10.2   INDIRECT CONTROL

Indirect social control refers to the relationships individuals have that positively or negatively influence their behavior. In other words, these relationships either encourage conformity to social norms or may encourage individuals to engage in deviance and crime. The most common indirect control relationships discussed by metal thieves involved other metal thieves and family.

10.2.1   Other Thieves

As established in Chap. 9, metal thieves tend to know other metal thieves. In fact, this knowledge of other metal thieves occurs more frequently within the taxonomy of Metal Thieves than in any other scrapper taxonomy. It is far more frequent for a metal thief to have peers who are also committing metal theft and therefore is less likely to experience positive indirect social control. Rather, a metal thief's violation of general societal and scrapper norms, values, and codes may be encouraged due to their peers.

This concept is magnified during criminal partnerships, which often occur, as demonstrated in Chap. 9. The significant influence partnerships have on indirect social controls is primarily due to the increased contact and influence, especially since many partnerships are formed through romantic relationships, between siblings, or emerge from long-standing friendships or while at work. The primary factor in each of these situations is the increased time of association. In other words, most criminal partnerships were also relationships that occurred during other associations such as work, family, and friendships. Metal thieves who worked in partnerships therefore spent a lot of time together. These intimate associations combined with criminal partnerships enhance indirect control, encouraging criminal behavior.

10.2.2   Family and Friends

The second common form of indirect social control is other family members and close friends who are not involved in metal theft. In general, family members and close friends, other than those involved in a criminal partnership, tended to have a positive indirect social impact on metal thieves. For example, Leo described one reason he quit stealing metal:

> *Leo*: You know I met this girl, and we got engaged and stuff, so at that point, it was scary, so that's why we had to get out of the business, and go out and get regular jobs...
> *Interviewer*: Is it still tempting to you?
> *Leo*: Oh yeah, definitely! Driving down the road you see a building, oh man that's a gold mine right there, you know, but I just can't do that to her (his fiancée). I want to be fair to her.

Similarly, Shane, a Subsistence Scrapper, described why he does not steal, saying, "Oh no! 'Cause [crime] it's in my name and I don't steal. Plus, my momma told me she would beat my fucking ass, and she died of cancer, and is in heaven looking down on me."

Conversely, a lack of family and close friends may also demonstrate the reduced indirect social controls available to some thieves. Chris, Daniel, David, and Robert described losing all or most of their close family members through death or because of shunning. For example, when discussing his family, Chris said in exasperation, "My dad's dead, my mom's dead, my family's dead, all of them!" Similarly, David related, "My family that is here has nothing to do with me, and the rest of them are dead, and my children right now don't have anything to do with me." In fact, thieves who lacked strong family connection seemed to take greater risks while stealing metal and seemed to be arrested more often.

Metal Thief Profile

Name: Robert
Taxonomy: Metal Thief
Location: Northeast
Age: 26
Race: White
Education: Associates
Employment: Full-Time—Mechanic
Experience: While waiting in line for several hours at a scrap yard the day before Christmas, I met Robert. We began to talk and he shared his long past with me. Nearly all of his family has passed away while he was young (17) and he turned to scrapping and metal theft to survive. He earned enough during those years through drug dealing and metal theft to pay for college. He became a mechanic and had a full-time job, yet continued to steal small amounts of metal from his company to recycle for extra income.

Most thieves who did have family and close friends withheld their criminal activities from them, presumably to avoid shame and indirect social control. For instance, James said about his family's knowledge, "They don't really know, I mean, it's just me and her (his wife), and we don't [tell] anybody." In the same way, John told his family about scrapping, but "nothing on the tobacco frames" he stole.

Several thieves discussed the negative impacts when their family discovered their criminal activity, usually in conjunction with an arrest. David and Matt both said, "I lost my wife," and Michael said, "I didn't tell anyone about it except people at work. When my wife found out, she divorced me; sucked, 16 years down the drain! After the arrest, people thought I was some type of murderer of something. I was an outcast. My life sucked." In all three of these cases, the awareness of criminal activity came after an arrest for a metal theft related crime. Moreover, David, Matt, and Michael all discussed the shame and disappointment at losing family over their exposed activity and indicated that the disappointment and disapproval by a wife or significant other, above other reasons, was why they are not engaging in criminal activity today.

Leo remarked that after he was arrested for stealing metal, it was his fiancée that kept him and his partner from returning to metal theft:

> *Interviewer:* Now is that (the arrest) what made you stop?
> *Leo:* I had a fiancée and she never really wanted me to do this, so when I got busted and went to jail it really, really scared her, so I gave it up.
> *Interviewer:* Yeah, what about your partner, has he been able to give it up?
> *Leo:* Yeah, we actually live together, me, my fiancée and my partner, we live together, and yeah, he gave it up too.

In Leo's case, his fiancée knew of these criminal actives before his arrest, but afterward the pressure placed on him by her was the impetus that pushed him and his partner out of metal theft permanently. In fact, Leo described how he has not returned to his lucrative metal theft activities, where he averaged, "$800 cash a day," only because of his relationship:

> *Leo:* That's why we had to get out of the business and go out and get regular jobs, which sucks because I'm not making as much money as I used to. But you know it makes her feel more comfortable, and I don't have to worry about someone knocking on my door waking me to take me to jail or have her waking up in the middle of the night, you know, getting a phone call that I got locked up… Right now I'm back to doing construction on a full-time basis but like I said in the winter time it's slow right now for me, so money is tight. Now I do think about scrapping all of the time, but I can't take that chance.
> *Interviewer:* Okay, so it's still tempting to you?
> *Leo:* Oh yeah, definitely! Driving down the road you see a building, oh man, that's a gold mine right there, you know, and I just can't do that to her. I want to be fair to her.

Clearly, indirect social control of close friends and relatives can have negative and positive impacts on the lives of metal thieves. Thieves may experience indirect social norms, values, and codes influencing their lives in a positive way, like when David, Matt, and Michael discussed how losing a wife or partner helped pull them from metal theft or as Leo described how he not only left metal theft due to a fiancée, but also resists its lure for her sake. Conversely, some thieves experienced negative indirect controls when family or other close friends encouraged them to commit a crime. However, indirect control of family and friends is not the only way thieves are influenced; internal control also affects them.

10.3 INTERNAL CONTROL AND NEUTRALIZATION

Internal social control refers to the beliefs, norms, and codes, or a sense of right and wrong, which a criminal possesses internally and that influences their behavior. It is believed that if individuals internalize societal norms and values, they will be less likely to commit a crime. However, when individuals fail to internalize these norms, crime may occur. Chapters 3 and 4 identified and discussed the universal norms, values, and codes for scrappers. Further, it was seen that the majority of metal thieves were once scrappers who likely internalized these social controls, yet as current metal thieves, failed to comply with these internal controls. Chapters 7 and 8 demonstrated that many metal thieves drift into criminal activity and often claim to enjoy metal theft. Thus, a paradox exists: if metal thieves internalize social—and especially scrapping—codes, values, and norms before theft, how do they drift into metal theft and enjoy the activity? One method commonly used by criminals who avoid the internal social controls of their conscience may be to neutralize their behavior.

Working under the theory and assumption of social control, Sykes and Matza (1957) developed the theory of neutralization to explain how criminals violate their internal beliefs and commit a crime. The theory of neutralization argues that moral codes are not replaced when a crime occurs, but rather that criminals drift from legal to illegal activities while justifying, denying, or neutralizing their behavior. Neutralization theory is supported by the present study in several ways, including suspending norms, justifying behavior, and appeals to a higher loyalty.

10.3.1   Suspending Norms

While the subculture of scrappers seems to have clearly defined values, beliefs, and codes under which they operate (see Chaps. 3 and 4), metal thieves do not demonstrate a cohesive concept of values, norms, and codes. Rather, they tend to suspend standards and codes that had previously been a part of their life when they were legal scrappers. For example, John, who scrapped legally, suspended his norms, values, and codes when "seeing tobacco frames and see[ing] how they've been sitting there for a few years and wasn't being used" and he decided to take the frames. When stealing the farming implement, he set aside the scrapping codes of *no theft* and *ask first*, which are common among most scrappers. When and how these norms are suspended for each metal thief varies. For example, before his foray into metal theft, John always asked before taking metal from a farm. However, when I asked John why he decided to steal metal, he struggled to provide an answer, saying, "We kind of ran out of scrap and went to taking tobacco frames and everything, it was the first time. I didn't really think it through; it was dumb, I reckon. I didn't think nothing through about it."

Metal Thief Profile

Name:	John
Taxonomy:	Metal Thief
Location:	Southeastern—Rural
Age:	Late 20s
Race:	White
Education:	Unknown
Employment:	Full-Time—Farm Labor
Experience:	I was connected with John through a former work acquaintance and we talked over the phone. John explained how he worked as a farm laborer and would often be assigned to clean out the farms he was working of the discarded cars, parts, wire fences, and so on, and to recycle the remaining metals. He described how they ran out of metals, so he and his metal theft partner started taking tobacco

frames. These are large metal trailers that tobacco is hung on when transporting it from the field to the barns. He tried to justify his actions by explaining that "$87 is hard to come by when you're getting little nitwit stuff out of the woods. You have to work four or five hours to get a trailer load and you go over there and you only have $60 or $70 worth. So it's just we just liked the spare money and we didn't really think it through and seeing tobacco frames and seeing how they've been sitting there for a few years and wasn't being used and we just kind of pulled in there and just took them." By taking what he considered old and unused metals, John really did not think that what he was doing was wrong. But as time went on, he and his partner got bolder and bolder and began stealing fairly new equipment until his partner was caught and "snitched" on him. He received probation for several years due to his thefts.

The drift toward crime seemed to be an especially critical component when suspending norms. As identified and discussed in Chap. 7, most metal thieves drifted from legal scrapping into metal theft. This drift, however, requires that thieves suspend the norms that had governed their lives as scrappers. Sometimes this occurred due to an immediate financial need, which is why Matt began stealing saying, "I can remember the first time I did it. I was scared to death, but I needed the money, I needed the money bad." Other thieves had someone encouraging them, and others simply said the temptation was more than they could bear. In all of these cases, there seemed to be an identified point at which the thief suspended their norms and committed theft. In fact, all thieves, if asked, remembered their first experience stealing metal.

Many thieves discussed how the thefts got easier are they continued stealing. Michael explained, "In the beginning, I didn't think it was wrong just because everyone else in the company did it. But deep down inside, I knew it was something that would be frowned upon." Yet Michael not only continued stealing, he brought his friend in to help him when the volume was more than he could handle. Leo described how "the first time I did it was scary but, you know, I wasn't really good at it, but over the years I got really, really good at it."

Despite the various reasons that thieves suspended their norms, one common feature echoed in all the comments. Once the norms had been suspended and there were positive results in the form of money from a successful theft, the norms were easier to suspend. For example, as Leo described it, "I did it the first time and made a substantial amount of money, and I was like what the hell, and it kind of went from there ... how did I feel when doing it? Not good, but better when I had the money." While it became easier for thieves to suspend their norms in most cases, they still believed their actions were wrong. For instance, Dustin explained how, while stealing and suspending his norms, he "felt very low, like a piece of shit." Likewise, Chris explained the regret he had over his norm suspension:

> *Interviewer:* Let me ask you, do you feel bad about doing this to someone's property?
> *Chris:* All the time. All the time. All the time.
> *Interviewer:* You still do it, you just feel bad?
> *Chris:* Ah hum, but I got to survive. I got to do something to survive. You know that is my thing; survive, survive. Wake up, do one thing that day, just focus on that day, just one day at a time. That is how I do it. Sometimes I live my life a minute at a time. You know what I am saying?
> *Interviewer:* Do you like living that way?
> *Chris:* No (very softly and sincerely), I don't like living this way. I hate it. It sucks.

For different reasons and in different ways thieves suspended the norms, values, codes, and beliefs that were common to legal scrappers. While the initial event leading to theft was a defining moment and the catalyst for future thefts, most thieves looked back on that instance with some regret. Many seemed to recognize that what they were doing was wrong, but as they continued to suspend the norms, the norms became easier to suspend, and they were less able to stop their criminal activities.

10.3.2   Justification

Simply suspending norms is not the only form of neutralization that metal thieves use. Many metal thieves also justify or deny their criminal behavior in an attempt to neutralize their action. For example, Dustin denied any injury when stealing air conditioners from apartment complexes, citing insurance absorbing the costs and justified other thefts from "the richer parts (of town)," indicating that he believed the wealthy could afford the theft of an air conditioner:

> I didn't mind taking from apartment buildings because the insurance company would take care of the replacement of the items. I never went into the 'hood and took anything, just from the richer parts; my thought process was I wasn't hurting the 'hood.

Moreover, Chris, who is homeless, denied responsibility and directly blamed his thefts on the government for failing to give him financial assistance and forcing him into the situation:

> *Chris:* If I got a disability check, I wouldn't be doing none of this.
> *Interviewer:* You don't think so? (Referring to the burglary and metal theft he had just committed)
> *Chris:* Nah, I'd be all right at $800 a month!

Chris was not the only one who blamed the government, or as Daniel referred to it, "the system." Leo represents several thieves who indicated the reasons they began stealing metal was due to a criminal history that made gaining employment difficult:

> You know … being a convicted felon, it's really hard for me to find a job. I'm not saying that it's not possible, but you know at that time I didn't want to work a 9- or 10-hour job and make $50 or $60 bucks a day, really. It just wasn't worth it to me.

Leo, who estimated that he earned over $250,000 in stolen copper sales within three years before his arrest, went on to explain how he justified stealing by denying that his actions injured others and condemning other metal thieves who, in his eyes, were worse than he was:

> *Leo:* A lot of scrappers they go to the foreclosed homes and stuff like that, and they get stuff like that out of abandoned buildings. I never went to homes or anything like that. I was kind of like; you know *I'm not a thief.* What I used to do is I would find places that had burnt down, like big industrial hotels or an insane asylum that were closed down, that they were just going to tear down anyway, and so I would go in there and take that, and I always wondered who is going to miss that? *It's still ille-*

gal; you get in trouble for it, [but] you get in more trouble going into a home and stealing copper than you would a big industrial abandoned building.

Interviewer: Okay, so part of the reason that you chose the industrial sites is because you didn't want to get in trouble, or did you have some personal, moral issue with the homes?

Leo: It was for the fact that I wouldn't get into as much trouble, [and] it was also a moral thing, you know. I'm a firm believer in karma, you know. If you scrap a house that is foreclosed, the bank can't sell the house for the money that they want because all of the plumbing is gone out of it. So, whatever I've displaced has to come out of their pocket ... it's kind of wrong to do that. But to scrap a building that's going to be torn down in a year or two, who's really going to miss it you know? I never stole anything from anyone that it would hurt their livelihood, you understand what I'm saying? I wouldn't go into a store and rob someone, I wouldn't rob a bank or a person, per se, because you're taking away from their livelihood, you know, so I do have a conscience. (Emphasis added)

Leo had developed an extensive system of denials of injury and neutralization that he internalized to justify his actions. While his thought process may appear to contradict itself, Leo seemed entirely at ease with his explanation and the justification for his actions, indicating that he had neutralized any negative feelings of guilt that he may have initially felt.

Leo was not alone in his sophisticated justification and neutralizations. Many of the same concepts were echoed by Eric, who was a member of a large theft ring that stole used wire from a major electrical corporation, earning the group well over half a million dollars:

I would always try to make it seem like, you know, we're not doing a bad thing. All this is doing (referring to the metal he stole) is going back to a giant corporation, to a CEO who is getting all the money, and it's just a little man trying to make a buck. I was just trying to justify what was going on in my own head.

10.3.3   Appealing to Higher Loyalty

The last method of neutralization involved a handful of metal thieves appealing to higher loyalties, in most cases to the virtues of recycling. For example, as James reached into a dumpster in the alleyway behind a business, he said, "It's going to the dumpster anyway, you just don't want us to have it" criticizing individuals who throw things away and try to prohibit him from taking them by putting up fences and "no trespassing" signs. Leo further commented, "So pretty much I'm recycling." Dustin mentioned, "You're also helping the environment," when discussing metal theft.

This form of neutralization often occurred when metal thieves were taking metal items they believed were going to be discarded anyway, such as metal from an abandoned building, a constructions site, or a dumpster. They tended to think that as long as they did not see a continued use or purpose for the metal item, it should be free for them to take. Sometimes this was little more than taking something from a dumpster, as James did while I talked with him. Other times the virtues of recycling were used to justify theft.

These three common neutralization techniques are difficult for those not involved in metal theft to understand and may appear contradictory. However, most of the thieves interviewed never seemed to notice the contradictions in their statements. For example, at no point did Leo seem to identify, consider, or believe that the owner of an abandoned building, from which he pillaged tens of thousands of dollars, stands to be at a financial loss similar to, if not more than, the owner of an abandoned residence that another thief has stripped. Rather, Leo and most other thieves justified their actions, denied the injury metal theft caused, and some even appealed to the higher virtues of recycling. In this way, they were able to neutralize the moral and ethical trepidation they had when violating the scrappers' codes and norms, which most of them had established before stealing.

10.4 Summary

There are three primary areas of social control: direct social control, indirect social control, and internal social control. All three kinds of controls influence metal thieves. Direct controls are the external risks associated with stealing metal. Most often, the direct controls thieves are affected by are dogs, cameras, witnesses, and law enforcement. These factors tend to influence how, when, and if a metal thief commits a crime but rarely entirely dissuade them from the theft.

Indirect social controls influence a thief's behavior largely based on the relationships a thief has with others. These relationships can be a positive influence, such as a condemning family member, or an encouraging influence, such as another metal thief or criminal partner. Social controls take the form of indirect controls that often involve a sense of guilt. The present study found that many metal thieves could neutralize any feelings of guilt, thereby lessening the impact of indirect control.

Social controls do have an influence, however moderate and fleeting, on the activities of metal thieves. However, due to neutralization techniques, the most significant social control aspect is the fear of punishment—direct control. Direct control is the area where prevention techniques are easily applied. For instance, increasing guardianship, including dogs and cameras on the property, observant witnesses, and police presence, has an impact on the activity of metal thieves. Finally, an extended absence from metal theft due to imprisonment may have the single greatest effect on controlling metal thieves.

REFERENCES

Nye, F. I. (1958). *Family relationships and delinquent behavior.* New York: Wiley.

Sykes, G. M., & Matza, D. (1957). Techniques of neutralization: A theory of delinquency. *American Sociological Review, 22*(6), 664–670.

CHAPTER 11

Conclusion and Implications for Public Policy

Within the last decade, the rates of metal theft in the United States have increased dramatically, causing significant damage to the built environment, individuals, businesses, and governments. Unfortunately, there have been very few empirical examinations of metal theft, therefore, the precise frequency and impact of metal theft are largely unknown. Moreover, to date, there is no known examination of metal thieves. This lack of data and understanding has been largely filled with anecdotal stories and media reports, which tend to highlight extreme cases of metal theft and create fallacious connections (see Chap. 6: The Drug Hyperbole). Regrettably, there is little to no understanding of who is committing metal theft, why, and how. This lack of data hampers any attempt to regulate, investigate, or reduce metal theft crimes. Researchers on metal theft have suggested that "more certain and more detailed results on the extent to which metal thieves plan their offending could be obtained using alternative methods such as offender interviews" (Ashby, Bowers, Borrion, & Fujiuama, 2014, p. 18).

This book represents the first of such efforts to understand scrappers and metal thieves through qualitative methods of observation, participation, and interviewing. The result is a rich understanding of metal thieves, their motivations, methods of learning, partnerships, theft methods and techniques, and the social controls that influence them. These findings provide a thorough understanding of metal thieves, offering guidance on future metal theft research and prevention methods.

© The Author(s) 2017   213
B.F. Stickle, *Metal Scrappers and Thieves*,
DOI 10.1007/978-3-319-57502-5_11

11.1   SIGNIFICANT FINDINGS

11.1.1   Scrappers versus Metal Thieves

One of the first difficulties in conducting an exploratory study of a population is identifying who is to be studied. In the present case, I was unable to locate literature that provided a comprehensive examination of the individuals involved in metal recycling. In fact, the colloquialism "scrapper" had never been defined. Moreover, a connection between scrappers and metal thieves was largely unknown, with many believing that scrappers are metal thieves.

The present study provides the first known definition of scrapping: "the act of regularly collecting fragmented, damaged, or discarded metal items, which are no longer useful or have not maintained their original value, to recycle them for financial profit."

By operationalizing the definition of a scrapper and examining the data collected, this research has identified five taxonomies: Subsistence Scrappers, Scrapping Professionals, Professionals who Scrap, Philanthropic Scrappers, and Metal Thieves. This categorization revealed that not only are there distinctive types of scrappers, but that metal thieves function through unique methods that preclude their inclusion into other categories of scrappers. Therefore, delineating scrappers separately from metal thieves is not only appropriate, it is necessary to establish the focus of any study on metal theft. In addition to the taxonomy, this study documented a well-established subculture amongst most scrappers containing norms, values, codes, and behaviors commonly shared across each of the first four types of scrappers.

11.1.2   Metal Thieves and Motivation

The present study reveals that metal thieves commonly drift from "legal" scrapping to theft, leaving behind many aspects of the scrapping subculture, and therefore should be considered two separate groups. Consequently, it is important to identify the demographics of metal thieves and understand what their motivations are, apart from scrappers.

Unfortunately, the majority of what is presumed in the literature about metal thieves is based on news reports and individuals (commonly law enforcement and industry professionals), anecdotal experiences, and perceptions. These stories highlight drug usage as a key factor in metal theft.

Based on the present study, however, many of these perceptions appear to be spurious. The metal theft population in the present study tended to be white, male, and young adult. However, several unusual demographics emerged. First, nearly 55% of the study sample had some college experience. Second, approximately 70% of metal thieves were employed full-time while stealing metal, which is significantly higher than the averages for scrappers. The findings also indicate that metal thieves are more educated than most other criminal populations and may be motivated to commit metal theft for different reasons.

There are several studies claiming that the motivation for metal thieves is connected with the increasing value of metals. This price-theft hypothesis has been empirically validated several times; however, the present study's findings were inconclusive. A portion of the metal thieves (approximately one-third) claimed to be affected by the price, but the magnitude of the effect and their actual behavior are unknown. A few thieves did not believe the price affected their theft habits, but the majority gave conflicting responses.

Drug usage and addiction have also been frequently cited in popular media and other official documents as a motivator for metal theft. While this theory has broad appeal among government officials, no empirical studies have evaluated the connection. The present study found that drug usage does not appear to be a significant factor in metal theft. Of the one-third who discussed using drugs, only a small portion indicated they were motived to steal to support a drug habit.

The common stereotypes of metal thieves as homeless, uneducated, and drug addicted—which is common in media and government reports—have not been sustained in the present study. Rather, metal thieves may be the exact opposite: educated, employed, experienced in a work field related to metal, and tend to steal for reasons other than drug usage. These findings are important, as the traditional stereotypes of metal thieves are inaccurate and may negatively affect the ability of law enforcement, legislatures, and recycling centers to develop prevention efforts.

11.1.3   Metal Thieves' Techniques and Methods

Several scholarly studies indicate that metal theft is more likely to occur in places containing large amounts of valuable metal when there is decreased guardianship and a high likelihood of repeat victimization, and that metal thieves often operate in a calculated manner. To evaluate these claims and other factors related to metal theft, I examined the data and developed three distinct stages of any metal theft: identifying a place, committing the theft, and selling the metal. Developing these stages allows the present study and future evaluations of metal theft to analyze metal theft in a clear and temporal order.

The present study confirmed the literature findings that point to the built environment as a major factor in metal theft. Specifically, places with low guardianship (e.g., abandoned buildings) were primary attractors for metal thieves. Moreover, if metal supplies were available in great quantities at such locations, criminals tended to return to the same place repeatedly to steal metal, thus supporting findings of frequent repeat victimization.

The techniques and methods used to accomplish the thefts varied to a great degree, often depending on the place of the theft, type and quantity of the metal, and the skill of the thief. Most metal thieves work together in partnership and often case a place to establish the presence of metal and evaluate risks of detection before the theft. These activities clearly indicate that metal thieves most often function in a calculated manner. This is not to say that some metal thieves will not steal when the occasion presents itself, but that nearly all of the metal thieves in the present study made calculated efforts to locate the metal, plan the theft, utilize tools to effect the theft quickly and without detection, and took steps to reduce the likelihood of exposure at recycling centers.

Interestingly, employment in an industry commonly working with or around metal is a significant factor in metal thefts. Specifically, the majority of thieves either had been employed in the past or were currently employed in areas such as construction or HVAC installation, as electricians or plumbers, and at recycling centers. The skills gained in these careers, along with the access to tools and awareness of the value of metals and—perhaps most important—knowledge of the places where metals are frequently found, mean that metal thieves with these work experiences are very skilled at stealing. Not only were metal thieves with these work credentials more efficient thieves, they also received less scrutiny when selling metal, as they were expected to have metal to sell routinely in the course of their work.

11.1.4   Metal Thieves' Social Controls

While social controls affecting metal thieves have not been previously addressed in the literature, the present study identified critical controls that may provide insight into future studies and crime prevention techniques. Metal thieves are influenced by all three kinds of controls: direct, indirect, and internal. The most important findings in the present study, however, are related to direct controls.

The direct controls discussed most by metal thieves were dogs, cameras, witnesses, and law enforcement. The presence of dogs, cameras, and witnesses tend to influence how, when, and if a metal thief commits a crime. In each instance, the power of the direct social control is the fear of discovery and an investigation by law enforcement with possible punishment in jail or prison.

Of the direct controls identified in this study, dogs were the most commonly avoided. Other factors, such as possible witnesses and cameras, caused increased caution, but rarely entirely circumvented the theft. These findings are significant for two reasons. First, it provides an evaluation of what control techniques have the greatest impact on metal thieves. Second, it provides clear areas for where metal theft prevention techniques can be easily applied.

11.2 POLICY IMPLICATIONS

The increase in metal theft has bewildered local governments and victims, frustrated police in investigation attempts, and perplexed legislatures on how to establish regulations and laws to stem the rising tide of metal thefts. Perhaps the most significant reason for the difficulty in responding to metal theft is due to the lack of knowledge. To date, very few empirical studies have been conducted on metal theft, and those which have are limited in scope (e.g., confined to one city) or do not focus on evaluating the thieves themselves. This void of knowledge had repeatedly been filled with anecdotal stories and wild assumptions about who metal thieves are and how they operate. These assumptions have only been proffered by media reports, law enforcement, and trade publications, but have been granted validity when repeated in official government reports. For example, the FBI and the DOE, among others, have repeatedly suggested that metal theft is the result of drug usage, specifically methamphetamine, without offering any evidence to support this claim. This study, however, indicates that drug usage is not a primary motivation for metal theft.

Further, many of the attempts to reduce metal theft have drawn a particular focus on recycling centers by mandating new regulations and laws, such as requiring government identification, requiring photos of the purchased metals, or demanding checks be mailed rather than cash paid to the recycler on site. In extreme cases, some states require recycling centers to hold the metal in its original conditions for several weeks to allow for possible investigation. Each of these efforts has significant unintended consequences for recyclers and the recycling industry. For example, requiring the recycling company to hold metal for days or even weeks before alteration or resale has a significant negative impact on space to store the material, which is often limited, and the daily fluctuating value of metals influences their profitability. Moreover, requiring a government identification to sell metal and requiring a check mailed to a home address harm Subsistence Scrappers, who generally do not steal metal, and may not have an ID, home address, or bank account. Moreover, these policies have encouraged thieves to have someone else sell the metal (e.g., a fence) rather than risk capture or identification himself or herself, to sell it in other jurisdictions, to damage or destroy items, and to develop other evasive tactics. Finally, individuals who are typically trusted at a recycling center (i.e., electricians, HVAC workers, etc.) may commit the majority of metal theft. These thieves are then able to circulate stolen and non-stolen material for sale at the recycling center with impunity. In fact, several laws aimed at curbing metal theft exempt plumbers, contractors, electricians, and others from many of the regulations.

Unfortunately, these legal and policy efforts have had little to no impact on metal thieves. This is evident in the continued rise of metal theft, unabated by new legislation and efforts to curtail the crime. It appears that these efforts are not sufficient, probably because legislators were not aware of the techniques and methods metal thieves use at each stage of the theft and, specifically, how stolen metal is sold to recycling centers. If the extent of the problem is unknown and the offenders are unstudied, it will be very difficult to examine the trends and effects of metal theft, let alone develop policies, strategies, and laws to combat it. As Luke Bennett concluded, there is "little research aimed at understanding the causes and consequences of, or developing the necessary policy response for, dealing with this problem" (2008, p. 183).

With these failures in mind, coupled with what has been learned through this study, the following policy suggestions are presented. These implications are discussed along natural division lines, law enforcement, legislation, and the recycling industry. However, many of these recommendations are closely linked and should be viewed as complementary to one another, rather than individually.

11.2.1   Implications for Law Enforcement

The first and most important policy implication for law enforcement is the need to establish a method of tracking instances of metal theft. Currently, few states have a commonly used, separate criminal charge or case identifier for metal theft incidents. Moreover, only a handful of cities is known to possess the ability to track metal theft. The result is that metal theft data is included with data on many other types of theft and it is either difficult or impossible to isolate. The lack of data significantly reduces the ability of the law enforcement community to identify, understand, and develop crime prevention techniques.

While law enforcement agencies can begin to track metal theft with their in-house system, agencies should consider encouraging lawmakers to establish a separate criminal code for metal theft. Creating a separate code will allow precise definitions to be established, specify targeted penalties, and result in the collection of accurate data on a statewide basis. Creating a different code or law about a particular type of crime is nothing new. In fact, many states have different criminal statutes for bicycle theft, damaging vending machines, and theft of mail matter, to name just a few. However, merely encouraging the creation of a separate statute is not enough. Law enforcement agencies should actively analyze the data they receive during collection. This data can be analyzed as a part of intelligence lead policing, problem-oriented policing, hot spot policing, and many other techniques currently used by law enforcement agencies. This data should identify factors associated with metal theft and allow for a proper response and innovative investigation techniques.

Moreover, law enforcement officers would benefit from a cursory knowledge of metal theft and metal thieves. The present study reveals that on several occasions metal thieves had contact with law enforcement, yet officers failed to observe or identify stolen metal or were easily convinced that the metal was obtained legally. Moreover, many officers may be under the impression that all metal theft is related to drugs, or that Subsistence Scrappers are thieves. These presumptions may lead them to fail to consider plumbers, for example, and emphasize their investigation on Professional Scrappers (who rarely steal) and thereby misappropriate resources and lead to fewer closed cases.

A brief training on many of the aspects of metal theft presented in this book would likely result in a significant increase in identifying stolen metal and prosecuting metal thieves. Training should cover four critical areas in which law enforcement needs to have greater awareness and understanding. First is an awareness of who is commonly involved in metal theft and how they function (e.g., in groups). Second, law enforcement officers should have an established understanding of the motivation of metal thieves (e.g., drug usage is unusual). Third, officers need to be aware of the methods and techniques of metal theft (e.g., using legitimate employment as a cover for theft). Lastly, the ability to ascertain the social controls likely to prohibit metal theft (i.e., dogs, cameras, witnesses) is an essential tool for crime prevention.

11.2.2   Implications for Legislators

Legislators in all 50 states have been active in altering or creating laws relating to metal theft. These legal changes include, among other things, increased criminal penalties for theft, restrictions on recycling centers (e.g., requiring a yard to hold the metal a certain number of days before sale or destruction), and prohibiting the sale of metals or specific types or forms of metals (e.g., manhole covers) without proper documentation. However, many of these efforts appear to have failed to hold back the wave of metal theft. This failure is due in large part to a lack of understanding of who the metal thieves are and how they operate, and may have unintended consequences for business owners and legal scrappers.

One of the most important policy changes legislators can implement involves defining metal theft. Many of the definitions of metal theft are inadequate because they are either too inclusive or too restrictive. Developing a firm definition of metal theft, such "the theft of item(s) for the value of the constituent metals" (Whiteacre et al., 2014) will enable legislatures to construct legislation aimed at deterring metal theft successfully. Included in this effort may be an additional charge for damage done to a building or structure that may not be traditionally considered "regular" theft cases.

For example, if a thief steals boxes of new, uninstalled electrical wiring at a construction site, the loss is directly related to the cost of the wire. If, however, after the electrical wiring was installed the thief ripped the wiring from behind the walls, there is a direct cost of replacing the wire, but also the indirect costs of repairing the drywall, paying for a new electrical inspection, and more. These factors should be considered when developing laws related to metal theft. Some states have adopted legislation aimed at increased penalties for the indirect damage caused by individual incidents of metal theft.

Finally, legislators should encourage the collection and dissemination of metal theft data to researchers, practitioners, law enforcement, and other stakeholders. This can be done by creating a task force to examine metal theft or assigning this goal to an already established group. It is important that a variety of persons participate in this task force; specifically, the group should include those that are most impacted by metal theft: victim advocacy groups, businesses, insurance agencies, utility and transportation companies, and law enforcement groups. In addition, the task force should include local or state researchers. The most important, but oft-neglected, participants are the recycling centers and recycling associations. Most industry leaders are eager to establish policies that will reduce stolen material entering their facilities and will likely welcome the opportunity to provide guidance and suggestions on reducing metal theft.

11.2.3   Implications for Recycling Centers

Since all stolen metal is sold to a recycling center, the greatest possibility of influencing the rate of metal theft is at these places. To varying degrees, efforts have been made to limit or require certain activities at recycling centers, primarily through legislation; however, these efforts have thus far proved ineffective. Based on the findings in the present study, the following policy suggestions would be more likely to influence the ability for thieves to sell stolen metal to recycling centers.

The first recommendation is that recycling centers require individuals who present credentials, such as trade licenses, to update them on an annual basis. This will ensure that the people recycling large amounts of metal have maintained their professional licensure. Moreover, in cases where a letter from an employer is provided as proof of legally acquiring the metal, recycling centers should require letters be updated on a quarterly or semi-annual basis and to contain detailed descriptions of the types and quantities of metal approved for recycling. For example, if a general contractor submits a letter on behalf of an employee stating that the employee is allowed to recycle scrap metal, the letter should be updated quarterly and should specify the type of metal (e.g., rebar, copper tubing).

Second, recycling center employees should receive training in identifying suspicious circumstances related to stolen metal. Training may be as simple as educating employees on the taxonomies of scrappers and metal thieves. This knowledge allows recycling center employees to focus on those who are most likely to be thieves (e.g., those employed in a metal related field), rather than on those who are unlikely to steal metal (e.g., Subsistence Scrappers). Additional training on the methods and circumstances thieves use to sell stolen metal would also be helpful. This training would include concepts such as altering or destroying metals, suspicious circumstances, or the use of fences to sell the metal for another person.

Third, recycling centers should participate in industry promoted metal theft prevention techniques. The Institute for Scrap Metal Recycling, for example, hosts a live and interactive website (http://www.scraptheftalert.com) for searching unusual metal sales and posting suspicious sales. This site is directly connected with law enforcement, if they choose to search it, and can help recover stolen items and reduce thefts. Unfortunately, this system is voluntary and reports only suspicious sales.

Finally, many states require recycling centers to record sales with photos and descriptions of metals purchased, as well as seller identity. In some cases, this data may be reported in real time to local police departments. This existing technology should be strengthened by linking recycling center computers together, in real time and across geographic boundaries, to provide instant notification of suspicious behavior. The present study identified some techniques thieves use to sell stolen metal (e.g., frequent small sales at multiple yards). These techniques could result in automatic notices at the time of purchase. For example, if the same person sells small amounts of metal at three different scrap yards within a few hours, the computer would automatically flag the activity and individual as suspicious. Then when the thief attempts to sell at the next recycling center, the computer would notify the company of the suspicious behavior. Similarly, recycling center employees could flag suspicious sales across the network, reducing the ability of thieves to avoid detection by visiting multiple recycling centers. Similar concepts have been successfully implemented with pharmacies in several states (see Blumenschein et al., 2010; Brady et al., 2014) to reduce prescription drug abuse.

11.3   SUMMARY

This exploratory study of a deviant subculture, scrappers, and a growing criminal activity, metal theft, is significant and groundbreaking in many ways. The information and rich understanding of scrappers and metal thieves garnered in this research are the first known qualitative study of the individuals involved in these activities. The present study has identified themes, developed concepts, operationalized definitions, and established the foundation for future research.

Before this study, criminology knew little about metal theft and almost nothing about how metal thieves thought and operated. This study has provided the first view of the unique and emerging crime of metal theft. It has corroborated a handful of empirical studies and refuted many of the popular media reports on metal theft. It has identified many recommendations for law enforcement personnel that will reduce the incidence of metal theft. With the knowledge base of this study, future research on metal theft and metal thieves will be able to advance the knowledge, present richer findings, and develop effective crime prevention strategies.

REFERENCES

Ashby, M. P., Bowers, K. J., Borrion, H., & Fujiuama, T. (2014). The when and where of an emerging crime type: The example of metal theft from the railway network of Great Britain. *Security Journal, 30*, 1–23.

Bennett, L. (2008). Assets under attack: Metal theft, the built environment and the dark side of the global recycling market. *Environmental Law and Management, 20*, 176–183.

Blumenschein, K., Fink, J. L., Freeman, P. R., James, K., Kirsch, K. L., Steinke, D. T., et al. (2010). *Kentucky all schedule prescription electronic reporting program (KASPER) evaluation team: Review of prescription drug monitoring programs in the United States.* Lexington, KY: University of Kentucky, College of Pharmacy, Department of Pharmacy Practice and Science Institute for Pharmaceutical Outcomes and Policy.

Brady, J. E., Wunsch, H., DiMaggio, C., Lang, B. H., Giglio, J. A. M. E. S., & Li, G. (2014). Prescription drug monitoring and dispensing of prescription opioids. *Public Health Reports, 129*(2), 139–147.

Whiteacre, K., Terheide, D., & Biggs, B. (2014). *Research brief: Metal thefts in Indianapolis October 1, 2011–September 30, 2013.* Indianapolis: University of Indianapolis Community Research Center.

INDEX

© The Author(s) 2017   225
B.F. Stickle, *Metal Scrappers and Thieves*,
DOI 10.1007/978-3-319-57502-5

The manufacturer's authorised representative in the EU is Springer
Nature Customer Service Centre GmbH, Europaplatz 3, 69115 Heidelberg,
Germany. If you have any concerns regarding our products, please
contact ProductSafety@springernature.com

Printed and bound by CPI Group (UK) Ltd, Croydon, CR0 4YY
23/04/2026
02095598-0002